TRANSFORMING TRAJECTORIES
FOR WOMEN OF COLOR
IN TECH

T0073909

Evelynn Hammonds, Valerie Taylor, and Rebekah Hutton, *Editors*

Committee on Addressing the Underrepresentation
of Women of Color in Tech

Board on Higher Education and Workforce

Policy and Global Affairs

A Consensus Study Report of

The National Academies of
SCIENCES · ENGINEERING · MEDICINE

THE NATIONAL ACADEMIES PRESS
Washington, DC
www.nap.edu

THE NATIONAL ACADEMIES PRESS   500 Fifth Street, NW   Washington, DC 20001

This activity was supported by contracts between the National Academy of Sciences, National Institute of Standards and Technology (SB134117CQ0017/1333ND20FNB100131), and National Science Foundation (CNS-1923245). Any opinions, findings, conclusions, or recommendations expressed in this publication do not necessarily reflect the views of any organization or agency that provided support for the project.

International Standard Book Number-13: 978-0-309-26897-4
International Standard Book Number-10: 0-309-26897-4
Digital Object Identifier: https://doi.org/10.17226/26345
Library of Congress Control Number: 2022932722

This publication is available from the National Academies Press, 500 Fifth Street, NW, Keck 360, Washington, DC 20001; (800) 624-6242 or (202) 334-3313; http://www.nap.edu.

Suggested citation: National Academies of Sciences, Engineering, and Medicine. 2022. *Transforming Trajectories for Women of Color in Tech*. Washington, DC: The National Academies Press. https://doi.org/10.17226/26345.

The National Academies of
SCIENCES · ENGINEERING · MEDICINE

The **National Academy of Sciences** was established in 1863 by an Act of Congress, signed by President Lincoln, as a private, nongovernmental institution to advise the nation on issues related to science and technology. Members are elected by their peers for outstanding contributions to research. Dr. Marcia McNutt is president.

The **National Academy of Engineering** was established in 1964 under the charter of the National Academy of Sciences to bring the practices of engineering to advising the nation. Members are elected by their peers for extraordinary contributions to engineering. Dr. John L. Anderson is president.

The **National Academy of Medicine** (formerly the Institute of Medicine) was established in 1970 under the charter of the National Academy of Sciences to advise the nation on medical and health issues. Members are elected by their peers for distinguished contributions to medicine and health. Dr. Victor J. Dzau is president.

The three Academies work together as the **National Academies of Sciences, Engineering, and Medicine** to provide independent, objective analysis and advice to the nation and conduct other activities to solve complex problems and inform public policy decisions. The National Academies also encourage education and research, recognize outstanding contributions to knowledge, and increase public understanding in matters of science, engineering, and medicine.

Learn more about the National Academies of Sciences, Engineering, and Medicine at **www.nationalacademies.org**.

The National Academies of
SCIENCES · ENGINEERING · MEDICINE

Consensus Study Reports published by the National Academies of Sciences, Engineering, and Medicine document the evidence-based consensus on the study's statement of task by an authoring committee of experts. Reports typically include findings, conclusions, and recommendations based on information gathered by the committee and the committee's deliberations. Each report has been subjected to a rigorous and independent peer-review process and it represents the position of the National Academies on the statement of task.

Proceedings published by the National Academies of Sciences, Engineering, and Medicine chronicle the presentations and discussions at a workshop, symposium, or other event convened by the National Academies. The statements and opinions contained in proceedings are those of the participants and are not endorsed by other participants, the planning committee, or the National Academies.

For information about other products and activities of the National Academies, please visit www.nationalacademies.org/about/whatwedo.

COMMITTEE ON ADDRESSING THE UNDERREPRESENTATION OF WOMEN OF COLOR IN TECH

EVELYNN M. HAMMONDS (*Chair*), Barbara Gutmann Rosenkrantz Professor of the History of Science, Professor of African and African American Studies, Harvard University

VALERIE TAYLOR (*Chair*), Director, Mathematics and Computer Science Division, Argonne National Laboratory; CEO and President, Center for Minorities and People with Disabilities in IT

GILDA BARABINO, President, Olin College of Engineering

SARITA E. BROWN, Co-founder and President, Excelencia in Education

JAMIKA D. BURGE, Director of Design Research, Capital One; Co-Founder of blackcomputeHER.org

FRANCES COLÓN, Senior Director, Center for American Progress

SARAH ECHOHAWK, Chief Executive Officer, American Indian Science and Engineering Society

ELENA FUENTES-AFFLICK, Professor of Pediatrics and Vice Dean for Academic Affairs, University of California, San Francisco

ANN QUIROZ GATES, Senior Vice Provost for Faculty Affairs and Director, Computing Alliance of Hispanic-Serving Institutions, The University of Texas at El Paso

SHAWNDRA HILL, Senior Lecturer, Columbia Business School, Columbia University

MARIA (MIA) ONG, Senior Research Scientist and Evaluator, TERC

MANUEL A. PÉREZ-QUIÑONES, Professor, Department of Software and Information Systems, University of North Carolina, Charlotte

KARL W. REID, Senior Vice Provost and Chief Inclusion Officer, Northeastern University

ALLISON SCOTT, Chief Executive Officer, Kapor Center for Social Impact

KIMBERLY A. SCOTT, Professor of Women and Gender Studies, Arizona State University

RAQUEL TAMEZ, Chief Executive Officer, Society of Hispanic Professional Engineers (*until January 2021*)

BRENDA DARDEN WILKERSON, President and Chief Executive Officer, AnitaB.org (*until November 2020*)

CYNTHIA WINSTON-PROCTOR, Professor of Psychology, Howard University; Principal, Winston Synergy, LLC

Study Staff

REBEKAH HUTTON, Program Officer and Study Director
ASHLEY BEAR, Acting Board Director

Preface

Being the first and only has been a characteristic of many women's lives in the science, technology, engineering, and mathematics (STEM) fields. The challenges they face in their academic and professional careers seem to resonate across sectors, yet for women of color, in particular, decades of efforts by academic institutions, industry, and government to create strategies for improving not only representation but also inclusion, belonging, and advancement are not moving the needle. Systemic racism, misogyny, ableism, ageism, and a multitude of other factors have perpetuated environments where women of color are often not able to fully use their talents in authentic ways. Increasing numbers of girls and women in tech without addressing these systemic factors will not produce sustainable change.

There is a critical need for institutions and organizations to take an intersectional approach—that takes into account how the intersection of race, gender, and economic disparities influences the experiences of women of color—when developing interventions aimed at improving equity, diversity, and inclusion in the vast number of STEM disciplines. Although this committee's task defines tech as computer science, computer and information science, information technology, and computer engineering, many of the findings and recommendations presented in this report have a broader applicability to transforming the experiences of women of color in tech across a broad range of sectors and contexts.

In much of the published research and data related to the representation of women in tech, the data for women of color have not been disaggregated, and the reported experiences of women do not reflect the experiences and representation of women of color, which can vary substantially from that of white women. Recent research described in this report demonstrates the importance of addressing the specific needs and particular challenges of women of color who are of African American, Latinx, Asian, American Indian, Alaska Native, Native Hawaiian, or other Pacific Islander descent. Within these groups, varied histories, cultures, communities, and support systems shape women's lived experiences in their academic and professional careers. Data that include these contexts will ensure that descriptions of the experiences of women truly reflect the experiences of *all* women. Given the widespread use of technology in all disciplines, ranging from the sciences to the humanities, analyses that take these nuanced experiences into account are critically important to transform the trajectories of women of color in technology such that they are engaged and driving technological innovations to produce robust and broadly applicable solutions.

In November 2017, the National Academies of Sciences, Engineering, and Medicine's Committee on Women in Science, Engineering and Medicine (CWSEM) organized a workshop on Women of Color in STEM attended by approximately 250 individuals. The workshop included two panels; the first one provided a historical perspective—identifying areas in which progress has been made to overcome barriers as well as barriers that still remain. The second panel focused on evidenced-based initiatives and programs to address these barriers. One of the outcomes of the November 2017 CWSEM Workshop on Women of Color in STEM was identification of the need for a consensus study focused on evidence-based initiatives and recommendations to increase the representation of women of color in tech. With funding from the National Science Foundation and the National Institute of Standards and Technology, the National Academies convened this committee to issue a consensus report on this topic, informed by a series of four regional workshops, published research literature, and other sources of data. The result is this report, with significant findings and recommendations that identify gaps in existing research that obscure the nature of challenges faced by women of color in tech, address systemic issues that negatively affect outcomes for women of color in tech, and provide guidance for transforming existing systems and implementing evidence-based policies and practices to increase the success of women of color in tech.

Persisting inequities such as opportunity gaps and the widening digital divide continue to reveal cracks in our society. Failure to address these issues will cause a generation of talent to be lost as a result of not having access to tools that allow them to access pathways into tech and to thrive in tech disciplines. It is critically important to disaggregate data using an intersectional approach to develop effective strategies for increasing the success of women of color in tech and meet these challenges head on. When policymakers, educators, and corporate leaders study issues related to women in tech, the term "women" must reflect the experiences of all women. We have the tools in place to effect the change we know is required to make a more equitable future. Now is the moment for true change and systemic transformation.

Evelynn Hammonds
Harvard University
Committee Co-Chair

Valerie Taylor
Argonne National Laboratory
Committee Co-Chair

Acknowledgments

This report would not have been possible without the contributions of many people. Special thanks go to the members of the committee who dedicated extensive time, expertise, and energy to the drafting of the report. The committee also thanks the members of the National Academies of Sciences, Engineering, and Medicine staff: Rebekah Hutton, Ashley Bear, and Priyanka Nalamada for their significant contributions to the report, Alex Helman for her early contributions to the committee's work, Crystal Grant for her contributions to helping organize the committee's workshops during her time as a Mirzayan Fellow, and Marquita Whiting and Abigail Harless for their administrative and logistical support of the committee.

The committee would like to thank Karin Matchett for her editing of the report as well as Heather Lavender, Nuria Jaumot-Pascual, Audrey Martínez-Gudapakkam, and Christina B. Silva for their work on the committee's commissioned literature review. We would also like to thank National Academies staff members who provided invaluable support throughout the project: Tom Rudin for his guidance and leadership; Anne Marie Houppert and Rebecca Morgan for their research support and fact checking; Adriana Courembis and Bardia Massoudkhan for their financial management assistance; Julie Eubank and Christopher King for their insights and guidance; Marilyn Baker and Erik Saari for their guidance through the report review process; and Amy Shifflette, Clair Woolley, and Holly Sten for their assistance with the final production of the report.

Many individuals volunteered significant time and effort to address and educate the committee during our four public workshops. The insights, perspectives, and personal experiences shared with the committee played an essential role in informing the committee's discussions and deliberations. We thank Stephanie

Adams, Carlotta M. Arthur, Twyla Baker, Sandra Begay, Kamau Bobb, Enobong "Anna" Branch, Mary Schmidt Campbell, Jennifer Carlson, Ashley Carpenter, Julie Carruthers, Nizhoni Chow-Garcia, Dilma Da Silva, Kathy DeerInWater, Andrea Delgado-Olson, Kaye Husbands Fealing, Dwana Franklin-Davis, Juan Gilbert, Raquel Hill, Evelyn Kent, Stephanie Lampkin, Bo Young Lee, Shirley Malcom, Marisela Martinez-Cola, Kyla McMullen, Carolina Huaranca Mendoza, Cherri Pancake, Melonie Parker, Alice Pawley, Denise Peck, Timothy Pinkston, Joan Reede, Dora Renaud, Monique Ross, Beena Sukumaran, Rati Thanawala, Rocío Medina van Nierop, Kenneth Walker, Bridgette Wallace, Gregory Walton, Gloria Washington, JeffriAnne Wilder, and Renee Wittemyer.

This Consensus Study Report was reviewed in draft form by individuals chosen for their diverse perspectives and technical expertise. The purpose of this independent review is to provide candid and critical comments that will assist the National Academies of Sciences, Engineering, and Medicine in making each published report as sound as possible and to ensure that it meets the institutional standards for quality, objectivity, evidence, and responsiveness to the study charge. The review comments and draft manuscript remain confidential to protect the integrity of the deliberative process.

We thank the following individuals for their review of this report: Lilia Abron, PEER Consultants, P.C.; Sandra Begay, Sandia National Laboratories; Quincy Brown, AnitaB.org; Gabriela Gonzalez, Louisiana State University; Vandana Janeja, University of Maryland, Baltimore County; Patty Lopez, Intel; Rati Thanawala, Harvard University; Roli Varma, University of New Mexico; and Renee Wittemyer, Pivotal Ventures.

Although the reviewers listed above provided many constructive comments and suggestions, they were not asked to endorse the conclusions or recommendations of this report nor did they see the final draft before its release. The review of this report was overseen by Ana P. Barros, University of Illinois at Urbana-Champaign, and Shirley M. Malcom, American Association for the Advancement of Science. They were responsible for making certain that an independent examination of this report was carried out in accordance with the standards of the National Academies and that all review comments were carefully considered. Responsibility for the final content rests entirely with the authoring committee and the National Academies.

Finally, the committee would like to thank the sponsors of this study for making this work possible. Funding for this study was provided by the National Science Foundation and the National Institute of Standards and Technology.

Contents

APPENDIXES

Boxes, Figures, and Tables

BOXES

FIGURES

TABLES

Summary

Demand for professionals in technology and computing fields is expected to increase substantially over the next decade, and increasing the number of women of color in tech will be critical to building and maintaining a competitive workforce. Women of color currently make up 39 percent of the female population in the United States and are projected to comprise the majority by 2060. In computing, women of color earn less than 10 percent of the bachelor's degrees and less than 5 percent of doctorates (McAlear et al., 2018). Despite years of efforts to increase the diversity of the tech workforce, women of color have remained underrepresented, and the numbers of some groups of women of color have even declined. Even in cases where some groups of women of color may have higher levels of representation, data show that they still face significant systemic challenges in advancing to positions of leadership. Research evidence suggests that structural and social barriers in tech education, the tech workforce, and venture capital investment disproportionately and negatively affect women of color.

Many efforts to increase the number of women in tech have focused on women more broadly rather than address the specific contexts of women of color. The lack of disaggregated data specific to women of color and subgroups of women of color has also impeded efforts to fully understand the causes and consequences of underrepresentation and develop effective strategies for increasing diversity, equity, and inclusion in tech at all levels. Recognizing the intersection of race, gender, and other social and cultural identities—and the ways in which those identities interact with existing systems—can provide insights and inform promising practices with the potential to increase the success of women of color in tech.

Although previous National Academies of Sciences, Engineering, and Medicine reports have addressed the underrepresentation of women in science, engineering, and medicine, none have focused on the unique experiences of women of color in tech disciplines. This report uses recent research as well as information obtained through four public information-gathering workshops to provide recommendations to a broad set of stakeholders within the tech ecosystem for increasing the recruitment, retention, and advancement of women of color. The committee's recommendations identify gaps in existing research that obscure the nature of challenges faced by women of color in tech, address systemic issues that negatively affect outcomes for women of color in tech, and provide guidance for transforming existing systems and implementing evidence-based policies and practices to increase the success of women of color in tech.

THE TASK

The committee was tasked by the National Science Foundation and the National Institute of Standards and Technology to convene four workshops and author a consensus study to examine strategies to improve the representation of women of color in tech. The committee's statement of task defines women of color as women who are African American, Hispanic, Latinx, American Indian, Asian American, Alaska Native, Native Hawaiian, or of other Pacific Islander descent. As previously noted, the committee recognizes that some subgroups within these populations may have better representation at different points in their academic and career trajectory. These demographic groups describe a broad range of ethnicities and geographic origins, and there is substantial diversity in the experiences and representation among and within groups of women of color.

The committee was tasked with (1) reviewing existing research literature and other resources to identify factors contributing to the underrepresentation of women of color in tech; (2) convening with experts from multiple regions of the United States to learn more about evidence-based, effective strategies for increasing recruitment, retention, and advancement of women of color in tech; (3) identifying factors that contribute to the success of women of color in tech; and (4) identifying, contextualizing, and disseminating recommendations for policy makers, academic institutions, employers, and other stakeholders (see Chapter 1 for the full statement of task).

For the purposes of this study, the committee's statement of task defines tech as computer sciences, computer and information science and support services, information technology, and computer engineering. The committee recognizes that the definition of tech as defined in its charge is narrower than the broad number of academic disciplines and careers that could be described as tech disciplines or careers. The committee made efforts to identify evidence and published research literature with particular attention to identification of sources directly related to the disciplines identified in the statement of task. However, the committee also recognizes the applicability and relevance of evidence from disciplines in tech defined more broadly, and has drawn upon published research as well as presentations from experts in science, technology, engineering, and mathematics (STEM), when appropriate, to further inform its deliberations and recommendations.

KEY CONCLUSIONS AND RECOMMENDATIONS

Based on the committee's evaluation of the research literature and other evidence, a key conclusion of the committee was that the experiences of women of color should inform the development of policies and practices intended to increase their representation in tech. Relatedly, the lack of disaggregated data poses a major challenge to understanding the nuanced and specific needs of different subgroups of women of color. Small sample sizes have frequently limited the collection of data specific to women of color in tech. The privacy concerns that arise in these small samples has also posed a challenge. While the committee recognizes the limitations of using small sample sizes and the limited ability to generalize findings to all women of color, the committee has approached its review of the evidence with a recognition of the fact that women of color are not a monolith and that there is value in exploring and understanding the experiences of subgroups of women of color. The committee concluded that use of appropriate qualitative data collection practices and other approaches that allow for the use of small sample sizes can inform the development of policies and practices based on the lived experiences of women of color to improve their representation, sense of belonging, and inclusion along their academic and career trajectories in tech.

The committee has provided targeted recommendations for future research and funding as well as recommendations for specific stakeholder groups in higher education, industry, government, and professional organizations. Recommendations appear at the end of each chapter in the report.

Recommendations for Future Research and Funding

More research, and more funding for research, should be dedicated to the following topics to significantly expand the knowledge base about how to better support and retain girls and women of color in technology and computing education and careers. Based on the committee's evaluation of research evidence and identification of gaps in research and funding, the following topics for future research and funding for K-12, higher education, and industry have been recommended.

K-12: Topics for Future Research and Funding

- Differences between girls of color and non-Hispanic white girls with regard to the digital divide, access to computer science courses, and quality of online learning

- Differences between girls of color and non-Hispanic white girls, and between women of color and non-Hispanic white women, with regard to educational experiences during the COVID-19 pandemic
- Intervention components that can positively impact the identity, confidence, interest, and aspirations of girls of color in tech and related fields, including counter-stereotypical role models, culturally relevant computing curricula, access to early childhood education that promotes culturally relevant socioemotional development, and diversity in the computer science teacher workforce

Higher Education: Topics for Future Research and Funding

- Impact of family support/encouragement and other early exposure experiences for women of color in tech and related fields
- How finances and financial aid (e.g., scholarships, loans and debt, salaries) impact women of color's entry into and persistence in tech and related fields
- Experiences of women of color at transition points throughout their academic career (e.g., from K-12 to higher education, from community college to four-year institutions, and from undergraduate to graduate education)
- Experiences of women of color in tech and related fields at the undergraduate and graduate levels at minority-serving institutions[1]
- Experiences of women of color in tech and related fields at technical colleges and community colleges
- Impact of peer mentoring on the success of women of color in tech
- Experiences of graduate students in tech and related fields who are women of color
- Experiences of women of color in tech in STEM and non-STEM community and counterspaces (i.e., safe spaces)

Workplace: Topics for Future Research and Funding

- Effective recruitment and hiring of women of color in tech and related fields
- Alternate pathway programs for women of color into tech careers
- Experiences of women of color in tech in STEM and non-STEM community and counterspaces (i.e., safe spaces)
- Award nominations and award receipt rates for women of color in tech

---

[1] In this report, a minority-serving institution refers to historically Black colleges and universities, Hispanic-serving institutions, tribal colleges and universities, and Asian American and Pacific Islander–serving institutions, collectively.

- The intrinsic qualities of women of color that contribute to persistence in tech and related careers
- Women of color in the tech workplace, specifically
  - how they enter the field (having a technical background vs. not having one)
  - promotion rates, experiences with employers, reasons for persistence or attrition
  - how finances (e.g., salaries, pay inequality) impact women of color's entry and persistence in tech and related fields

Recommendations: Challenging Assumptions Around the Recruitment, Retention, and Advancement of Women of Color in Higher Education

The committee offers the following recommendations regarding the recruitment, retention, and advancement of women of color in higher education.

RECOMMENDATION 3-1. To foster continuous pathways for women of color in higher education, institutions at the departmental, college, and university levels should promote the collection of empirical qualitative and quantitative data that disaggregate the recruitment and graduation experiences of students, the recruitment and promotion and tenure trajectories of all faculty, and ascension to leadership positions for women of color.

These data should be used to inform the design and implementation of the following processes, but not limited to

- culturally responsive review of promotion and tenure guidelines and academic review processes to ensure that the qualitative and quantitative research produced by women of color in tech is equally valued at the departmental, college, and university levels.
- collection, analysis, and presentation of disaggregated data of tech departments and college environments to institutional leaders. Information regarding the individuals who constitute research teams, laboratories, faculty service committees, and doctoral committees could be used to determine whether one group is disproportionately receiving opportunities or assuming more invisible labor. These data should also include the social categories (e.g., race/ethnicity, gender, socioeconomic status) of decision makers at departmental, college, and university levels in order to understand how power operates as an intersectional concept.
- a reward system sustained by computing and other technology-related departments and college environments which demonstrate ongoing levels of success recruiting, retaining, and maintaining an inclusive context for

women of color in tech. Both disaggregated qualitative and quantitative data could be used to present cases that illustrate effective strategies.

RECOMMENDATION 3-2. Institutions of higher education should collect and analyze disaggregated qualitative data to document the voices of women of color in tech and the narrated experiences of those who work with women of color that demonstrate how women of color fare in technology and computing courses as they navigate higher education at various levels.

To accomplish this, leaders in higher education, such as provosts, deans, and department heads, should use these data as the basis for their decisions for developing, sourcing, and evaluating initiatives for students and faculty who are women of color. Leaders in higher education should

- Regularly review and interpret these narrative data as barometers for measuring progress toward diversity, equity, and inclusion goals, and
- Identify and adopt best practices from institutions that have successfully recruited and retained women of color in tech.

RECOMMENDATION 3-3. Higher education leaders should widen recruitment efforts to identify women of color candidates to join their computer science, computer engineering, and other tech departments as students and faculty, with increased consideration of those from two-year community colleges and minority-serving institutions, and should develop retention strategies focused on supporting these students and faculty during transitions to their institutions.

Strategies should include the following:

- Developing partnerships with two-year community colleges and minority-serving institutions to identify and recruit tech students and graduates who are women of color.
- Increase access to higher education by integrating financial assistance programs with recruitment and retention strategies that target undergraduate and graduate students who are women of color.
- Providing increased social supports for incoming tech students and faculty who are women of color, such as orientations, professional development, career coaching, and peer mentoring. Individuals who provide this support should be required to maintain ongoing, regular training in culturally responsive education, racial awareness, and intersectionality.

Recommendations: Increasing Recruitment, Retention, and Advancement of Women of Color in the Tech Industry

The committee offers the following recommendations for increasing the recruitment, retention, and advancement of women of color in the tech industry.

> **RECOMMENDATION 4-1.** To enhance the accuracy of data reporting, tech companies should disaggregate employment data by tech and non-tech positions, job titles, gender, and race/ethnicity—with particular attention to the intersection of race/ethnicity and gender—and make those data publicly available. Reports should include information about trends in recruitment, retention, and advancement of women of color.

Although some companies have been hesitant to disclose EEO-1 data to the public, many other companies recently began releasing company-wide demographic data over the past few years. However, the majority of these reports classify women as a single underrepresented group despite vastly different trends among women from different racial and ethnic backgrounds. There remains a need for further transparency in order to fully understand the employment landscape of women of color in the tech industry. Demographic data play a critical role in measuring progress, along with identifying areas where additional resources are needed to improve recruitment, retention, and advancement; benchmarking; and creating strategic plans for improving diversity, equity, and inclusion.

Companies, organizations, and researchers also need data on recruitment demographics, promotion rates, and attrition (both involuntary and voluntary exits) in order to identify inequities and remove structural and systemic barriers that contribute to women of color leaving high-tech positions. Transparency in reporting will promote accountability. Without these types of changes, it is unlikely that the tech sector will be able to reduce racial bias and discrimination.

> **RECOMMENDATION 4-2.** Companies and organizations working within the tech sector should create pathways for women of color into leadership positions and create positions for diversity, equity, and inclusion professionals that are part of executive leadership.

Creating a diverse and inclusive organizational culture starts with leaders both in individual companies and across the industry who recognize their essential role in shaping organization culture and diversity, equity, and inclusion (DEI) priorities. Diversity is a business imperative that requires the oversight and attention of senior executive management. Diversity, equity, and inclusion professionals within a company should have sufficient financial and human resources to support organizational goals in order to successfully implement research-based best practices with well-defined goals. They should also have direct access to other members of the leadership team, an opportunity to report on the status of progress, and they should be able to demonstrate measurable success in their role.

Increasing the number of women of color in leadership positions will improve equity in tech by building industry leadership that reflects the identities of the customers and communities the industry serves. It is important to note that continuity of leadership, sustained implementation of best practices, and consistency of metrics for assessing the success of DEI efforts are factors that can reduce the negative effects of frequent organizational change (e.g., lack of promotion or the need to rebuild credibility with teams, customers, and partners) and improve outcomes for women of color as they progress in their careers.

Women of color bring a wealth of skills, abilities, networks, and other cultural capital to the workplace. Evidence shows that companies with more diversity in leadership outperform companies with less diverse leaders. In addition, cultivating more leaders who are women of color can improve innovation, increase recruitment of other women of color, and, in the long term, improve the diversity of the tech industry's talent pool.

> **RECOMMENDATION 4-3. Tech companies, with the assistance of a neutral central organization, should initiate an ongoing cross-sector coalition with each other as well as other stakeholders such as academic institutions—especially minority-serving institutions (e.g., historically Black colleges or universities, Hispanic-serving institutions, and tribal colleges and universities)—and professional societies. This collective would allow member organizations and institutions to connect with each other with the goal of supporting current and future women of color in tech and promoting effective recruitment, retention, and advancement strategies for women of color in tech across all entities.**

To create new solutions to improve diversity, equity, and inclusion, a collective approach to problem solving that facilitates the development of partnerships across the tech ecosystem could be a successful way to address the underrepresentation of women of color in tech. As other sectors increasingly utilize and develop new technologies, the tech sector will continue to expand and evolve. Although some collectives of tech companies already exist, a cross-sector, collective approach to strategic planning implemented in collaboration with a neutral, well-resourced central organization will help industry, higher education institutions, and other stakeholders (e.g., organizations working to create alternative pathways into tech and policy making) to increase accountability, share data, and leverage their strengths to develop strategies for improving policies and practices that improve outcomes for women of color as they transition from higher education into the workforce and as they advance in their careers in tech.

RECOMMENDATION 4-4. Tech companies should expand employment options that promote work-life balance such as remote work, flexible work hours, parental and other family leave, and career counseling as a strategy to improve retention and advancement and expand recruitment of women of color.

There are increased opportunities for recruiting and retaining a diverse workforce when companies implement practices that facilitate balance between work and home life. Although many companies within the tech sector have implemented flexible work policies, employees' opportunities to advance may sometimes be limited when they take full advantage of such policies. Although flexible work policies that promote work-life balancing have been shown to benefit both men and women, evidence shows that women—and women of color, in particular—are more negatively affected by the absence of these types of policies. Women shoulder a disproportionate burden of household management, childcare, and other caregiving. Research shows that implementing flexible work policies for all employees fosters more equitable participation in the workforce and increases opportunities for retention and advancement. Flexible work policies such as remote work may also be a valuable recruitment tool for attracting new employees who are women of color and allowing them the option to remain in geographic regions where they have easier access to family, community, and other support networks and resources.

Recommendations: The Role of Government in Addressing the Underrepresentation of Women of Color in Tech

The committee offers the following set of recommendations related to the role of government in addressing the underrepresentation of women of color in tech based on the findings presented in chapter 5.

RECOMMENDATION 5-1. Government efforts aimed at addressing the underrepresentation of particular groups in tech should intentionally account for intersectionality.

5-1 A. Any legislation aimed at addressing issues of underrepresentation in STEM and in tech should take an intersectional approach that considers the unique experiences of women of multiple marginalized identities (as described in Box 5-1).

5-1 B. Government efforts calling for data collection related to groups underrepresented in STEM and in tech should clearly indicate that such data be disaggregated by race/ethnicity and gender (to the extent possible given the need to protect anonymity of individuals) and should require qualitative as well as quantitative data collection, especially when the numbers are small enough that qualitative data would provide more meaningful information.

5-1 C. Program solicitations and descriptions at federal agencies should be explicit in directing prospective grantees to take an intersectional approach.

History demonstrates that unless policies, practices, programs, and individuals embrace an intersectional approach to promote diversity, equity, and inclusion in our institutions, women of color will not benefit from these efforts. The committee found that both legislative language and federal program solicitations related to diversity, equity, and inclusion were inconsistent in calling for an intersectional approach.

> **RECOMMENDATION 5-2. Federal agencies should submit to Congress an overview of their programs that support the recruitment, retention, and advancement of women of color in tech with their annual budget request as the National Science Foundation currently does in its Summary Table on Programs to Broaden Participation (see Table 5-1). If agencies do not create such annual reports voluntarily, Congress should mandate that agencies do so.**

In general, information about existing federal efforts aimed at supporting women of color in tech is widely dispersed and inconsistently distributed on various agencies' websites. The highly distributed nature of this information makes it challenging to gain a complete understanding and an accurate record of these investments. One notable exception is the National Science Foundation (NSF), whose annual budget request to Congress provides an annual compilation of the agency's efforts to support broadening participation.

> **RECOMMENDATION 5-3: To promote transparency and accountability, Congress should amend section 709e of the Civil Rights Act of 1964 to require public release of EEO-1 workforce demographic data by companies, which would include those that are the recipients of government contracts supported by taxpayer dollars.**

Research demonstrates that increasing transparency and accountability in diversity, equity, and inclusion efforts can yield tangible positive impacts. Recognizing the importance of data collection, transparency, and accountability, many investors have called upon tech companies (many of which are recipients of large government contracts) to be more transparent about the composition of their workforce by publicly releasing the EEO-1 demographic data that most companies are required to provide to the Equal Employment Opportunity Commission annually. In the committee's view, the public should be afforded the opportunity to hold these government contractors—some of which are the recipients of billions of taxpayer dollars—accountable for making progress toward their stated missions to improve the diversity of their workforce.

> **RECOMMENDATION 5-4. Federal agencies should incentivize grantee institutions' efforts to improve diversity, equity, and inclusion through accountability measures.**

5-4 A. Prospective grantees' plans to promote diversity, equity, and inclusion should be reviewed by review panels and agency personnel and should be a determining factor in awarding or renewing funding to an institution, in addition to technical merit. Grantees should include a description of the impact of their efforts to promote diversity, equity, and inclusion in annual reports and requests for funding renewals.

5-4 B. Federal agencies should invest in programs that incentivize institutional efforts to take a culturally responsive, intersectional approach in promoting diversity, equity, and inclusion in tech through award and recognition programs, such as the SEA Change effort led by the American Association of the Advancement of Science, which is currently funded by the National Science Foundation, the National Institutes of Health, and a number of private foundations.

5-4 C. Federal agencies should carry out periodic "equity audits" for grantee institutions to ensure that the institution is working in good faith to take an intersectional approach to address gender and racial disparities in recruitment, retention, and advancement.

- Institutions could be electronically flagged by the funding agency for an equity audit after a certain length of time or amount of funding is reached.
- An evaluation of the representation of women of color among leadership and academic success of women of color disaggregated by department should be included in such an audit.
- Equity audits should include a statement from institutions to account for the particular institutional context, geography, resource limitations, and mission and hold that institution accountable within this context. The statement should also account for progress over time in improving the representation and experiences of underrepresented groups in science, engineering, and medicine and should indicate remedial or other planned actions to improve the findings of the audit.

- The equity audit should result in a public-facing report made available on the agency's website.[2]

5-4 D. Federal agencies should consider institutional and individual researchers' efforts to support greater equity, diversity, and inclusion as part of the proposal compliance, review, and award process. To reduce additional administrative burdens, agencies could work within existing proposal requirements to accomplish this goal. For example, NSF could revise the guidance to grantees on its broader impact statements and the National Science Board could carry out a review of past NSF awards to determine how the NSF directorates have accounted for gender equity, diversity, and inclusion among the metrics evaluated in proposals submitted to NSF.

Federal agencies can play a powerful role in holding grantees accountable and by incentivizing action at institutions. If these recommendations are implemented with an intentional focus on intersectionality, it is the committee's opinion that they could be a positive force for holding institutions accountable for working in good faith to address the underrepresentation of women of color in tech education and careers.

> **RECOMMENDATION 5-5. Professional organizations and associations that represent the scientific and tech community (e.g., the Association for Computing Machinery, the Association for Computing Machinery, the Institute of Electrical and Electronics Engineers, the American Association for the Advancement of Science) should consider partnering with organizations that are committed to dismantling structural racism, such as the NAACP, National Urban League, LULAC, UnidosUS, Native American Rights Fund, United Negro College Fund, and National Congress of American Indians, to extend their sphere of influence and expand their outreach to policymakers on issues related to diversity, equity, and inclusion in tech fields.**

Strategic partnerships that extend an organization's sphere of influence are key to promoting policy change. There are examples in science and education policy in which meaningful policy change has grown out of partnerships and coordinated advocacy efforts. Advocacy coalition frameworks and specific case study examples could serve as models to stakeholders, such as scientific and engineering professional societies, that are working to advocate for improving the recruitment, retention, and advancement of women of color in tech (Weible, 2017; Weible and Ingold, 2018; Weber, 2019). The committee sees an opportunity

---

[2] This recommendation is also put forth in NASEM (2020).

for scientific and engineering professional societies (e.g., American Association for the Advancement of Science, American Physical Society, American Chemical Society, National Society of Black Engineers) and higher education associations (e.g., the Association of American Universities) that engage in advocacy for science and for diversity, equity, and inclusion in STEM, to form strategic partnerships with influential organizations that have worked for many years to address structural racism and sexism and which have a great deal of influence with government institutions.

Recommendations: Alternative Pathways for Women of Color in Tech and the Role of Professional Societies

The recommendations that follow address the roles of academia, community organizations, industry, federal agencies, and professional societies in increasing the number of women in tech through education of K-12 students and retraining programs for adults.

RECOMMENDATION 6-1. Industry and funding agencies should invest in expansion of certification and training programs for women of color that are delivered by community-based organizations to scale their capacity to recruit and prepare a greater number of women of color in tech. These investments should expand opportunities for apprenticeships and people seeking to (re)enter the tech workforce.

The low number of women of color in tech positions who have not received a bachelor's degree (Table 6-1) and who earn certificates (Figures 6-1 and 6-2) demonstrates that women of color are not taking sufficient advantage of alternative pathways into tech careers. Recently, there has been significant interest in reskilling in computing-related areas among non-computer science majors (NASEM, 2018; NAS, NAE, and IOM, 2005). Re-entry programs provide a substantial opportunity for women who stepped away from the workplace for family reasons and seek re-entry, perhaps in a career they may not have previously considered.

Professional preparation of women of color can be an integral component of an organization's diversity, equity, and inclusion strategy. Dedicated efforts in areas of national need, such as artificial intelligence, cybersecurity, and data analytics, can provide entry into tech fields and provide women of color the appropriate knowledge, skills, and abilities that can lead to progressively more advanced roles.

RECOMMENDATION 6-2. Funding agencies should invest in programs that provide scholarships to Native female students who pursue a graduate program in a computing-related field and commit to teach at a tribal college or university for the length of the scholarship.

Tribal colleges and universities are an important entry point into technology and computing fields for Native female students; however, there are not many tribal colleges and universities that offer a bachelor's or master's degree; most are similar to a community college (Varma, 2009a, 2009b). While these institutions provide more curricula aligned to the culture of American Indians and Alaska Natives (Ambler, 2002), institutions are influenced by the structural and geographical challenges experienced on reservations where they are located—for example, high unemployment rates, low per-capita income, lack of qualified instructors, and hard-to-reach locations (Varma, 2009a, 2009b). These factors represent barriers for prospective faculty to teach technology at these institutions. Only 12 out of 35 tribal colleges and universities offer career pathways in computing.

Providing incentives to acquire the credentials needed to teach at a tribal college or university could leverage a common desire of women of color to give back to their community. This desire connects to Carlone and Johnson's (2007) concept of the altruist scientist, whose scientific identity is tied to altruistic values connected to science as the means to improve people's lives. The literature demonstrates that many women of color consider altruistic values as an intrinsic part of their identity as scientists and seek to give back by supporting their communities, mentoring or serving as role models, and supporting those who are like them in some way, such as sharing their same gender and/or race/ethnicity or being interested in similar fields (Agbenyega, 2018; Foster, 2016; Herling, 2011; Hodari et al., 2014, 2015, 2016; Lyon, 2013; Rodriguez, 2015; Skervin, 2015; Thomas, 2016).

The NSF Cybercorps® Scholarship for Service[3] program provides a model for increasing the number of computing programs offered at tribal colleges and universities. Scholarship for Service offers scholarships to students who pursue a post-baccalaureate degree in cybersecurity who commit to working for the federal, state, local, tribal, or territorial government, or a federally funded research and development center, after graduation for a period equal to the duration of the scholarship. Such a program could provide financial support for Native female students to seek a post-baccalaureate degree and give back to their community by becoming an instructor at a tribal college or university in a computing-related field.

---

[3] For more information see https://beta.nsf.gov/funding/opportunities/cybercorps-scholarship-service-sfs-0.

RECOMMENDATION 6-3. Higher education administrators should incentivize technology and computing-related departments to accept tech-related certification and digital badges, and should provide well-defined pathways for women of color and others from technology training programs offered by community colleges, industry, and especially community-based organizations toward earning associates, undergraduate, and graduate degrees in tech fields.

Industry-based training programs represent new pathways for employees and others to earn advanced degrees in technology. The programs provide educational benefits through partnerships with higher education institutions such as Northeastern University, one of the first institutions to offer workplace badges for academic credit. It is not clear if women of color are taking advantage of these emerging pathways, perhaps because of the high barrier to entry into industry positions. On the other hand, community-based technology training programs, particularly those that target women of color, provide supportive environments for women to gain information technology skills and earn certifications and badges. These programs tailor their recruitment messaging, instruction, and wrap-around services and provide ongoing support for their alumna.

RECOMMENDATION 6-4. Professional societies should create programs and/or initiatives directed at developing additional pathways that advance women of color in tech. These programs should have a strong evaluation component to demonstrate impact and provide recommendations for scaling successful models. Programming should include certification and badging options defined collaboratively with, and recognized by, industry and academic partners. Moreover, professional societies should be intentional about diversifying their internal leadership.

Professional societies support the development of standards and are positioned to "design and promote change, including through publications, policy statements, meetings, committees, lectureships, and awards" (NAS, NAE, and IOM, 2005). Furthermore, these societies often offer educational and informational resources and can offer support to students who are interested in educational and career opportunities in a specific discipline (NAS, NAE, and IOM, 2005; Morris and Washington, 2017). Unfortunately, this influence is not often exercised as effectively as it could or should be concerning increasing diversity. While professional societies may episodically focus on their outreach to women and individuals from underrepresented groups, they often experience little success in increasing engagement or participation.

The real key to engaging and broadening participation is designing programs and initiatives that are shaped by and for the groups they purport to target (Morris and Washington, 2017). A review of the longstanding professional societies that support individuals in one or more of the STEM fields revealed no programs or initiatives focused specifically on creating pathways or advancing women of color in tech, though some do have programs or initiatives for women in tech and/or STEM, and/or people of color in tech and/or STEM. Moreover, the leadership of professional societies rarely reflects the diversity of the future workforce. Even among professional societies that specifically serve people of color, there are very few with any programming or initiatives solely for women of color in tech.

REFERENCES

Agbenyega, E. T. B. 2018. "We are fighters": Exploring how Latinas use various forms of capital as they strive for success in STEM. PhD dissertation, Temple University.

Ambler, M. 2002. Sustaining our home, determining our destiny. *Tribal College* 13(3):8. https://tribalcollegejournal.org/sustaining-home-determining-destiny/.

Carlone, H. B., and A. Johnson. 2007. Understanding the science experiences of successful women of color: Science identity as an analytic lens. *Journal of Research in Science Teaching* 44(8):1187-1218.

Foster, C. 2016. Hybrid spaces for traditional culture and engineering: A narrative exploration of Native American women as agents of change. PhD dissertation. Arizona State University, Tempe, Arizona.

Herling, L. 2011. *Hispanic women overcoming deterrents to computer science: A phenomenological study*. Vermillion, SD: University of South Dakota.

Hodari, A. K., M. Ong, L. T. Ko, and R. R. Kachchaf. 2014. New enactments of mentoring and activism: U.S. women of color in computing education and careers. Proceedings of the 10th Annual Conference on International Computing Education Research. Pp. 83-90.

Hodari, A. K., M. Ong, L. T. Ko, and J. Smith. 2015. Enabling courage: Agentic strategies of women of color in computing. In *2015 Research in Equity and Sustained Participation in Engineering, Computing, and Technology (RESPECT)*, Institute of Electrical and Electronics Engineers. Pp. 1-7. doi: 10.1109/RESPECT.2015.7296497.

Hodari, A. K., M. Ong, L. T. Ko, and J. M. Smith. 2016. Enacting agency: The strategies of women of color in computing. *Computing in Science and Engineering* 18(3):58-68.

Lyon, L. A. 2013. Sociocultural influences on undergraduate women's entry into a computer science major. PhD dissertation. University of Washington, Seattle, WA.

McAlear, F., A. Scott, K. Scott, and S. Weiss. 2018. *Women and girls of color in computing*. Data brief. Kapor Center. https://www.wocincomputing.org/#data-brief.

Morris, V. R., and T. M. Washington. 2017. The role of professional societies in STEM diversity. *Journal of the National Technical Association* 87(1):22-31.

NASEM (National Academies of Sciences, Engineering, and Medicine). 2018. *Sexual harassment of women: Climate, culture, and consequences in academic sciences, engineering, and medicine*. Washington, DC: The National Academies Press. https://doi.org/10.17226/24994.

NASEM. 2020. *Promising practices for addressing the underrepresentation of women in science, engineering, and medicine: Opening doors*. Washington, DC: The National Academies Press. https://doi.org/10.17226/25585.

NAS, NAE, and IOM (National Academy of Sciences, National Academy of Engineering, and Institute of Medicine). 2005. Facilitating interdisciplinary research. Washington, DC: The National Academies Press. https://doi.org/10.17226/11153.

Rodriguez, S. 2015. Las mujeres in the STEM pipeline: How Latina college students who persist in STEM majors develop and sustain their science identities. PhD dissertation. University of Texas, Austin.

Skervin, A. 2015. *Success factors for women of color information technology leaders in corporate America*. Minneapolis, MN: Walden University.

Varma, R. 2009a. Attracting Native Americans to computing. *Communications of the ACM* 52(8): 137-140.

Varma, R. 2009b. Bridging the digital divide: Computing in tribal colleges and universities. *Journal of Women and Minorities in Science and Engineering* 15(1):39-52.

Thomas, S. S. 2016. An examination of the factors that influence African American females to pursue postsecondary and secondary information communications technology education. PhD dissertation. Texas A&M University, College Station. http://hdl.handle.net/1969.1/156994.

Weber, R. 2019. Keys to successful advocacy: The role and value of coalitions. Naylor: Association Advisor. https://www.naylor.com/associationadviser/successful-advocacy-coalitions/.

Weible, C. 2017. The advocacy coalition framework. International Public Policy Association. https://www.ippapublicpolicy.org/teaching-ressource/the-advocacy-coalition-framework/7.

Weible, C., and K. Ingold. 2018. Why advocacy coalitions matter and how to think about them. Policy & Politics. 46(2):325-345. https://doi.org/10.1332/030557318X15230061739399.

1

Introduction

Improving the representation of women of color in science, technology, engineering, and mathematics (STEM) is a national imperative. A 2011 report from the President's Council of Advisors on Science and Technology projected that 1 million more STEM professionals are needed by 2030 to maintain our nation's global competitiveness. In addition, a 2016 report from the Intel Corporation and Dahlberg Global Development Advisors, *Decoding Diversity: The Financial and Economic Returns in Tech*, suggests that increasing racial and ethnic diversity and ensuring full representation of gender diversity in the U.S. technology workforce has the potential to add $470 to $570 billion in new value to the U.S. tech industry. To achieve this ambitious national goal, we must draw upon all available talent, including the recruitment and retention of more women of color in STEM (PCAST, 2012), and address longstanding inequities and exclusionary practices that impede our ability to utilize the talents of those who are committed to STEM careers. Although tech can refer to a number of academic disciplines and careers across a wide variety of sectors within STEM, the committee's statement of task limits its focus to computer science, computer and information science and support services, information technology, and computer engineering; however, the committee has drawn upon research literature and other evidence from related STEM disciplines to inform its discussion of and recommendations for increasing the representation of women of color in tech levels of their academic and professional careers (Box 1-1).

In 2009, the National Science Foundation's Committee on Equal Opportunities in Science and Engineering hosted a "Mini Symposium on Women of Color in STEM," chaired by Evelynn Hammonds of Harvard University and Mia Ong of the Technical Education Research Center. The report recommended that the

---

Box 1-1
Women of Color

The committee has been tasked with presenting evidence and providing recommendations that recognize the specific needs and varying experiences of women of color throughout their academic and career trajectories. The committee's statement of task defines women of color as women who are African American, Hispanic, Latinx, American Indian, Asian American, Alaska Native, Native Hawaiian, or of other Pacific Islander descent. In much of the literature related to increasing the representation of women of color in tech and other STEM disciplines, Asian American women are often excluded; however, this demographic group describes a broad range of ethnicities and geographic origins. Although some subgroups of Asian American women may be overrepresented among STEM degree earners, only very small numbers are advancing to the ranks of full professor and serving in university leadership (e.g., deans or university presidents), and Asian American women remain underrepresented on corporate boards of trustees and among managers in industry and government. Similarly, the committee recognizes that there is substantial diversity within other groups of women of color included in this study. Subgroups within these larger categories may have not only differing experiences (e.g., the experiences of Black women who are descendants of enslaved Africans vs. the experiences of first-generation Black women of African origin) but also differing levels of representation. However, a key finding of the committee was the lack of disaggregated data specific to the groups of women that are the focus of this report—frequently as a result of small sample sizes and privacy concerns. The committee's recommendations call for increased data collection and transparency in reporting that can inform the design and implementation of practices that improve the experiences among and within groups of women of color.

National Science Foundation and Congress support "workshops and conferences for critical stakeholders (i.e., professional societies, university department chairs in STEM, honor societies) to discuss preparation of women of color for employment and share evidence-based best practices." The present report is a direct response to the 2009 report.

The present study was also informed by a series of reports from the National Academies of Sciences, Engineering, and Medicine and other research organizations, including the following:

- *Promising Practices for Addressing the Underrepresentation of Women in Science, Engineering, and Medicine: Opening Doors* (NASEM, 2020).
- *Barriers and Opportunities for 2-Year and 4-Year STEM Degrees: Systemic Change to Support Students' Diverse Pathways* (NASEM, 2016).
- *Double Jeopardy? Gender Bias Against Women of Color in Science* (Williams, Phillips, and Hall, 2014).

- *Women in Tech: The Facts* (Ashcraft et al., 2016).
- *Seeking Solutions: Maximizing American Talent by Advancing Women of Color in Academia* (NASEM, 2013).
- *Expanding Underrepresented Minority Participation: America's Science and Technology Talent at the Crossroads* (NRC, 2011).
- *Gender Differences at Critical Transitions in the Careers of Science, Engineering, and Mathematics Faculty* (NRC, 2010).
- *Inside the Double Bind: A Synthesis of Empirical Research on Undergraduate and Graduate Women of Color in Science, Technology, Engineering, and Mathematics* (Ong et al., 2011).
- *The Mini-Symposium on Women of Color in Science, Technology, Engineering, and Mathematics (STEM): A Summary of Events, Findings, and Suggestions* (Ong, 2010).

Over the last two decades, the proportion of underrepresented women of color (African American, Latinx, American Indian, Alaska Native, Native Hawaiian, and other Pacific Islander) who receive degrees at the bachelor's, master's, and doctoral levels has more than doubled (NSF, 2013, 2017). However, despite this important achievement, women of color remain significantly underrepresented relative to the national population, with the exception of some subgroups of Asian American women. The underrepresentation of women of color in STEM fields is especially pronounced in the tech sector (Ashcraft et al., 2016). Major tech companies such as Apple and Dell report that only 3 percent and 4 percent of their employees are African American/Black women, respectively, while companies like Facebook, Google, Intel, Microsoft, Yahoo, and LinkedIn report that only 1 percent of their employees or fewer are African American/Black women (Dillon et al., 2015). The underrepresentation of these women in tech careers is strongly influenced by the small number of women of color who pursue academic majors in tech areas such as computer science and computer and information science. According to recent surveys from the Computing Research Association and the Association for Computing Machinery, U.S. universities are experiencing burgeoning enrollments in tech-related majors (Camp et al., 2017a, 2017b), a trend that is projected to continue given the growing impact of computer science on nearly every sector of business, academic disciplines, and most aspects of modern life. A significant proportion of the growth in enrollment for students of color has come from minority-serving institutions. Despite this overall increase in enrollments, the data also demonstrate a decline in enrollments for Black/African American students. In addition, enrollments for women and students of color decrease as course level increases (Camp et al., 2017b). Camp and colleagues (2017b) noted that efforts to increase enrollments may reduce diversity but reported a significant correlation between academic units that took actions to assist with diversity goals and higher proportions of women and students of color. They also reported that only 14.9 percent of academic units specifically chose actions with consideration of reducing impact of those actions on diversity, and only 11.4 percent decided against actions because of potential impacts on diversity. The current expansion in student enrollment in tech-related education is an opportunity to engage and empower institutions to learn from the past and support a culture of inclusivity. They can do so by considering diversity, equity, and inclusion in the development of strategies to recruit and retain women of color to ensure that students who receive these degrees are more ethnically and gender diverse than they have been historically.

Women of color at all levels of their academic or professional careers—including those who have advanced—who work in tech commonly experience feelings of isolation (i.e., feelings of invisibility or hypervisibility), macro- and microaggressions, and a sense of "not belonging" (Ong et al., 2011). In addition, women of color are often excluded from informal professional networks and are more negatively affected when institutions or organizations do not have career-life balance policies that address gender differences in regard to pressure associated with having a family (Kachchaf et al., 2015). These experiences of bias and exclusion that question women's competence, contributions, ambition, and leadership can culminate in women doubting their abilities and experiencing what is often described as "imposter syndrome"—a phenomenon that is particularly prevalent in biased and inhospitable organizational cultures (Tulshyan and Burey, 2021).

A 2014 study based on interviews with 60 women of color working in STEM fields (20 each of Latinx, Asian American, and Black women) and an online survey of 557 women in STEM (both women of color and white women) found pervasive gender bias (Williams, 2014). One hundred percent of the women interviewed reported experiencing some form of gender bias (Williams, 2014), including prove-it-again bias (Eagly and Mladinic, 1994; Foschi, 1996, 2000), the tightrope bias (Cuddy et al., 2004; Fiske, 1999), the maternal wall bias (Cuddy et al, 2004; Correll et al., 2007), and/or the tug-of-war bias (Derks et al., 2011a, 2011b) (Box 1-2). The type of bias varied by ethnicity. Black women were more likely to report the prove-it-again bias, Asian women students benefited from the stereotype that Asians are "good at science," Latinas reported being pressured by colleagues to do administrative support work for their male colleagues, and nearly half of Black and Latina women reported regularly being mistaken as custodial or administrative staff. Women of color who work in science and engineering may experience isolation (Williams and Dempsey, 2014), racial stereotypes, accent discrimination, and a lack of role models, effective mentors, and professional networks (Kachchaf et al., 2015).

Educational institutions and employers have implemented many interventions to improve the representation of women in STEM, but these efforts tend to primarily benefit white women (Ong et al., 2011). Thus, it is essential to adopt an intersectional approach to the issue of diversity in tech to account for the

---

BOX 1-2
Common Forms of Bias Experienced by Women in STEM

The following terminology was originally used by Williams and colleagues (2014) and Williams and Dempsey (2014). The works of other scholars have confirmed these phenomena.

- **Prove-it-again bias**: In which women have to provide more evidence of competence than men in order to be seen as equally competent (Eagly and Mladinic, 1994; Foschi, 1996; Ong et al., 2011).
- **The tightrope bias**: In which women have to walk a "tightrope" between being seen as too feminine to be competent or too masculine to be likeable (Cuddy et al., 2004; Foschi, 2000).
- **The maternal wall bias**: In which women are assumed to experience declines in their work commitment and competence after they have children (Correll et al., 2007; Fiske, 1999; Foschi, 2000).
- **The tug-of-war bias**: In which gender bias against women fuels conflict among women, such that women distance themselves from other women (Derks et al., 2011a, 2011b).

---

complex, cumulative ways in which multiple forms of discrimination (e.g., racism and sexism) intersect in the experiences of women of color.

INTERSECTIONALITY

Intersectionality is an important component to discussions and analysis of the experiences of women of color. The first few subsections provide an overview of intersectionality and related concepts, followed by a description of the committee's approach. For more than 30 years, Kimberlé Crenshaw, a lawyer, civil rights intellectual leader, and legal scholar, has used critical race theory to interpret the law. She coined the term "intersectionality" to describe the ways that multiple forms of inequality can be compounded to create obstacles that do not align with conventional ways of thinking about social advocacy. In 1989, Crenshaw published *Demarginalizing the Intersection of Race and Sex*, in which she described Black women's experiences of discrimination as being at the intersection of gender and race, which, as axes of analysis, are not mutually exclusive. Her work demonstrated the shortcomings of prevailing antidiscrimination law and policy as related to Black women's experiences, and, subsequently, to the experiences of *any* woman of color or people experiencing multiple forms of discrimination. In particular, Crenshaw demonstrated that women of color were excluded from legal narratives about racial justice, which largely focused on Black men's experiences, and gender equality, which largely focused on white women's experiences.

Crenshaw's intersectional lens provides the conceptual context for a broader exploration of systemic inequalities that are perpetuated by the systems in which women of color live, learn, and work.

Over the years, the application of intersectionality to research and practice has exploded across multidisciplinary scholarship. Patricia Hill Collins (2015) noted that these variations expose a growing challenge: While intersectionality represents an opportunity to define the complexities of gender and racial disparities as well as other identity-based inequities (such as age and ability), *how* we conceptualize and understand intersectionality is far from clear or consistent. As intersectionality has grown in popularity and as an accepted field of study, some parts of its narrative are better understood, while other efforts to address intersectionality dilute or weaken its original intent (Collins, 2015). For example, it is generally understood that women of color experience unique forms of gender and racial discrimination, but what is often lost in translation is the fact that women of color are not a monolithic group. Within the populations of groups that are the focus of this committee's report, there can be substantial variation in experiences and representation. As a whole, women of color require distinct and unique supports, relative to the inequitable power and social systems they experience; however, these supports must be informed by the varying experiences and challenges faced by different groups of women of color.

The Double Bind and Intersectionality

In December 1975, the American Association for the Advancement of Science convened a two-day workshop chaired by Jewel Plummer Cobb, a member of the National Science Board and advisor to American Association for the Advancement of Science. Thirty Black, Mexican American, Native American, and Puerto Rican women were invited to participate, representing various fields of science, engineering, medicine, and dentistry and coming from a variety of educational and work experiences and generational and geographic backgrounds. The workshop participants discussed the experiences of countless women who had been consistently excluded as contributors to and scholars of the scientific community. In her preface to the workshop report, Plummer Cobb wrote:

> Although this group of women came from the "pure" and applied sciences, with a wide range of ages and experiences and diverse backgrounds and cultures, we shared a common bond; and a special and warm sense of sisterhood sprang from this. Generation gaps did not divide us, nor did our varied vocations, nor our cultural diversity. The common ties were those of the double oppression of sex and race or ethnicity plus the third oppression in the chosen career, science (Malcom et al., 1976, p. ix).

The report of this workshop described this double oppression as the "double bind," the systematic biases—based on racism and sexism—that women of color experience in the professional scientific community (Malcom et al., 1976). These findings—which provided insight into the "differentness" that women of color experienced across educational (pre-collegiate and collegiate), familial, cultural, and societal dimensions—described the experiences of the 30 women who attended the workshop. Forty-five years later, the same experiences echo within the STEM community of women of color, particularly in the contemporary contexts of educational and workforce development.

The double bind contends that men of color and white women must understand two dimensions related to the experiences and realities of women of color: first, that enormous and unique demands are commonly placed on them and, second, that racial and gender identities cannot be separated from their lived experiences. Intersectionality acknowledges the interlocking nature of racial and gender discrimination that women of color experience and underscores the roles of power and privilege in causing and, in some cases, enforcing discrimination at multiple levels.

Systems of Power and Intersectionality

Collins (2015) highlighted variation in scholars' definitions of intersectionality and defines attention to power and social inequalities as a common thread. Moreover, she argued that categories of group membership are best understood from an orientation of relational terms because categories of group membership "underlie and shape intersecting systems of power; the power relations of racism and sexism, for example, are interrelated" (Collins, 2015, p. 14).

To inform solutions intended to increase the representation and success of women of color in tech, it is critical to consider systems of power and oppression that affect individuals' experiences. Winston and Winston (2012) highlighted several characteristics of systems of power within racialized societies such as the United States. They asserted that power systems have important attributes: (1) structure, form, and stability over time; (2) uniform and identifiable patterns; and (3) systems may be "theoretically neutral" as in Brazil or systems may partly be specified in law as in the southern United States, contrasted with customary practices as exemplified in de facto segregation in northern states. Critical institutions become the effective operational structures of such discriminatory systems (e.g., education, housing, and the health care system).

The systems theory most relevant to intersectionality is critical race theory. Many scholars trace the origins of critical race theory to legal scholars' analyses of critical legal studies. These scholars argued that critical legal studies restricted the legal ability to analyze racial injustice in the law and within critical institutions because the studies did not adequately consider race and racism (Bell, 1985, 1987; Crenshaw, 1989; Crenshaw et al., 1995; Delgado, 1988; Winston, 1991). To address this deficiency, scholars developed a jurisprudence that accounts for the role of racism in U.S. law, asserting that the failure to explain how systems of power promote structural inequalities that systematically discriminate against individuals, based on their racial group membership, maintains and perpetuates hegemony and the status quo. Similarly, legal scholars in the late 1980s and 1990s described the inadequacies of the U.S. laws at that time to account for both racial and gender discrimination simultaneously experienced by women of color in the workplace (Crenshaw, 1989; Crenshaw et al., 1995; Winston, 1991). Thus, it is important that the use of the term intersectionality in the context of women of color in tech recognizes its roots within legal studies and U.S. law.

The Committee's Perspective on Intersectionality

This committee used intersectionality as an analytic framework to interpret evidence about the underrepresentation of women of color in tech. As such, the committee recommends solutions that account for the interconnectedness of the multiplicative experiences of sexism, ageism, and racism encountered by women of color in tech within the organizations where they learn and work. The committee also considered it important to unpack the term "women of color" in order to better understand the experiences of Asian, Black, Indigenous, and Latinx women. The experience of people in each group—and within each group—is steeped in gender and racial inequities that intersect differently and thus require interventions and policies that acknowledge those differences. In a similar way, women of color within each group have distinct and diverse backgrounds, cultures, and histories within the United States and throughout the world. Collectively, women of color share similar experiences, but their communities are not monoliths. The committee also believes that it is important to challenge assumptions about the meaning of "underrepresented," as it does not imply an impoverished background or an inherent inability to excel academically or professionally. As a result, the committee has taken the opportunity, when possible, to avoid the use of words like "minority" and "underrepresented" to describe individuals who have been historically excluded, instead describing the people we seek to support and understand clearly, rather than identifying them in statistical terms.

There are also systemic differences in how women of color need to be supported with respect to access to equity and excellence in education and the workplace; therefore, the committee's definition of intersectionality extends beyond describing racial and gender group membership (i.e., the intersection of race and gender) to incorporate dimensions of oppression, power, and privilege that plague the tech environments in which women of color learn and work. In terms of education, systemic differences include funding for educational equipment, identification of role models, and an inclusive curriculum. For the workplace, it is important to acknowledge power structures that may be barriers to success or to develop specialized programs that support the unique experiences of women of color. Often these differences are influenced by power and privilege to create barriers to the entry and persistence of women of color, which can result in low numbers of women of color in tech at all levels.

Throughout the report, the committee calls for the scientific community to adopt an intentional intersectional approach in its efforts to study the experiences of women of color in tech. Too often, the methodologies the scientific establishment relies upon, which may apply well in some situations, do not account for intersectionality and therefore do not yield critically important information that is needed to better understand the experiences of women of color in these fields or identify the factors that are affecting their attrition or lack of advancement in these fields. Slaton and Pawley (2018) articulated this well:

> Because of the small numbers of women of color in engineering, analysts deem it methodologically necessary to aggregate all women together even when participants' experiences differ by race, or to aggregate all African American participants together even when their experiences differ by gender, even though this methodological decision results in losing not just nuance but entire lived experiences of engineering education. The idea of what may comprise "representative" experience (in the statistical sense) takes precedent, and may discourage researchers' engagement with the idea of intersectional identities. In any inquiry where sheer numbers dictate viable populations for study, we risk dismissing curiosity about the forms of identity (for example, along lines of sexuality, dis/ability, nationality, or age) that are associated with the most severe underrepresentation. Meaningful patterns of student participation in engineering education are seen to reside only in studies above a certain scale, as prevailing evidentiary standards determine what may and may not be subject to study. In other words, the logics of acceptable methods function to ensure that some stories are never studied because there aren't "enough" of them, even though their scarcity is precisely what makes the subjects critical to study.

In keeping with the rationale described previously, much of the research reviewed by the committee in this report is from projects that use qualitative research methods with small sample sizes. The purpose of such qualitative research is to gain an in-depth understanding of a phenomenon—often centered around answering why and how the phenomenon is experienced and the meanings it holds for people—which can be done with small samples of participants (APA, 2019; Creswell, 2013; Glaser and Strauss, 1967). Research based on qualitative methods, such as interview work, aims to understand phenomena by creating categories from the data and then analyzing relationships between categories, while paying attention to the lived experiences of the research participants (Charmaz, 1990, 2006). The committee recognizes the limitations of data using small sample sizes and the ability to generalize the findings to women of color at scale; however, qualitative research conducted within the field of psychology uses standards for quality more aligned with its nature, such as methodological integrity, rather than standards typical in quantitative research such as generalizability, validity, and reliability (APA, 2019; Levitt et al., 2017). The committee has approached its review of evidence with a recognition of the value of understanding the unique experiences of individual women of color. In addition, the committee reviewed literature related to STEM fields more broadly, when applicable, to inform its understanding of trends in the experiences of women of color in STEM that also affect women of color in tech.

THE STUDY CHARGE AND COMMITTEE'S APPROACH

The National Academies of Sciences, Engineering, and Medicine named a diverse, multidisciplinary committee to fulfill the statement of task in Box 1-3.

BOX 1-3
Statement of Task

An ad hoc committee of the National Academies of Sciences, Engineering, and Medicine will examine strategies to improve representation of women of color in technology and issue a consensus report informed by a series of four regional workshops. The committee will

1. Leverage the existing literature, and other resources as appropriate, to identify the factors that contribute to the underrepresentation and success of women of color in technology (i.e., computer science (CS), computer and information science and support services (CIS), information technology (IT), and computer engineering);
2. Organize four workshops to bring together the regional community to solicit evidence-based, effective strategies for increasing the success of women of color in tech;
3. Identify, contextualize, and disseminate recommendations that policy makers, academic institutions, employers, and other stakeholders can use to have positive impacts on the recruitment, retention, and advancement of women of color in tech;
4. Catalyze the building of communities of practice devoted to increasing the success of women of color in tech; and
5. Empower policy makers, academic institutions, employers, and other stakeholders with evidence-based practices for improving the success of women of color in tech.

Based on the content of four regional workshops and other information gathered, the committee will issue a consensus report that (a) identifies the factors contributing to the success of women of color in tech based upon the existing literature and other sources and (b) provides recommendations that policy makers, academic institutions, employers, and other stakeholders can use to have a positive impacts on the recruitment, retention, and advancement of women of color in tech. The committee may also produce rapporteur-authored proceedings in brief for one or more of the workshops.

As previously noted in this chapter, the committee acknowledges that the tech disciplines, as defined in its charge, are a subset of the broad number of academic disciplines and careers that are solidly based in technology. The committee made efforts to identify evidence and published research literature, with particular attention to identification of sources directly related to the disciplines identified in the statement of task. However, the committee also recognizes the applicability and relevance of evidence from disciplines in tech defined more broadly, and has drawn upon published research as well as presentations from experts in STEM, when appropriate, to further inform its deliberations and recommendations.

With the goal of reducing the impact of bias and discrimination, the committee hosted a series of four workshops (one in-person, and three virtual) that focused on systems-level changes, including the identification of practices, strategies, and policies for the academic, industry, and government sectors to change the culture, climate, norms, and values in tech fields. The workshops focused on intersectionality in order to ensure that the strategies and models would specifically address the multiple categories of identity that apply to women of color. The committee's scope included a broad range of career stages and tech sectors, including students and faculty in higher education and professionals in industry and government agencies.

The barriers and challenges encountered by women of color in STEM fields are well documented, but our knowledge of effective strategies to remove barriers is very limited.

The workshops were intended to achieve the following objectives:

- Articulate the evidence-based programs, models, and practices that academic institutions, employers, and individuals can implement to have a positive impact on the retention, recruitment, and advancement of women of color in tech.
- Engage and empower stakeholders with evidence-based practices to improve equity and diversity in these fields.

The committee also commissioned a literature review to summarize the body of empirical research on the topic of women of color in tech.

The committee took steps toward building a community of practice, bringing together a network including women and men who participated in the committee-sponsored workshops, to address the underrepresentation of women of color in tech and create novel research-practitioner partnerships. The committee's in-person workshop in February 2020 had over 100 attendees, and the attendance for the virtual workshops in April, May, and June 2020 ranged between 358 and 578 attendees.

The purpose of the network is to catalyze changes in the climate, culture, norms, and values of academic departments, business teams, government agencies and laboratories, and other entities in tech and engineering fields. This report summarizes the findings derived from the regional workshops, the commissioned literature review, and the committee's discussions.

OVERVIEW OF THE REPORT

In the chapters that follow, the committee reviews the state of knowledge on the factors that contribute to the underrepresentation of women of color in tech education and careers in industry and higher education. The report recommends the actions that a range of stakeholders could take to support the improved recruitment, retention, and advancement of women of color in these fields. While the report's recommendations are primarily focused on women of color in tech, many apply to women of color in STEM academic disciplines and careers defined more broadly.

Chapter 2 presents an overview of current knowledge on the structural, social, and psychological barriers faced by women of color in tech education and the workplace, and presents an overview of strengths and assets operating at the individual, community, and institutional levels that help women of color persist and succeed in tech and related fields. In Chapter 3, the committee focuses specifically on institutions of higher education and describes assumptions frequently perpetuated by leaders in higher education about the barriers to recruitment, retention, and advancement of women of color in tech in their institutions. The committee then draws on data, both statistical and empirical, to recognize and challenge these assumptions. Chapter 4 describes the unique challenges in industry settings and the social and environmental factors, both inside and outside of tech, that have the potential to increase the recruitment, retention, and advancement of women of color in tech careers. Chapter 5 reviews government efforts to support equity, diversity, and inclusion in tech and points out the many ways in which government efforts have often failed to take an intersectional approach in national programs, policies, and initiatives. This chapter also describes research on the importance of transparency and accountability in promoting change and highlights opportunities for Congress and federal agencies to take steps to promote transparency and accountability among tech companies and in the government itself. Chapter 6 describes alternative pathways that have opened up to meet the growing demand to facilitate entry into the tech workforce, such as employer-offered training, certification courses offered by two-year and four-year colleges, community-based and nonprofit organizations, apprenticeship and re-entry programs, and digital badging, and the role of professional societies in supporting women of color in tech.

REFERENCES

APA (American Psychological Association). 2019. *The APA publication manual* (7th edition). Washington, DC: American Psychological Association.

Ashcraft, C., B. McLain, and E. Eger. 2016. *Women in tech: The facts.* Boulder: National Center for Women and Technology.

Bell, D. A. 1985. The Supreme Court 1984 term foreword: The civil rights chronicles. *Harvard Law Review*, 99,4.

Bell, D. A. 1987. *And we are not saved: The elusive quest for racial justice.* New York: Basic Books.

Camp, T., W. R. Adrion, B. Bizot, S. Davidson, M. Hall, S. Hambrusch, E. Walker, and S. Zweben. 2017a. Generation CS: The growth of computer science. *ACM Inroads* 8(2):44-50.

Camp, T., W. R. Adrion, B. Bizot, S. Davidson, M. Hall, S. Hambrusch, E. Walker, and S. Zweben. 2017b. Generation CS: The mixed news on diversity and the enrollment surge. *ACM Inroads* 8(3):36-42.

Charmaz, K. 1990. 'Discovering' chronic illness: using grounded theory. *Social Science & Medicine,* 30(11), 1161-1172.

Charmaz, K. 2006. *Constructing grounded theory: A practical guide through qualitative analysis.* Thousand Oaks, CA: Sage Publications.

Collins, P. 2015. Intersectionality's definitional dilemmas. *Review of Sociology* 41:1-20.

Correll, S. J., S. Benard, and I. Paik. 2007. Getting a job: Is there a motherhood penalty? *American Journal of Sociology* 112(5):1297-1339.

Crenshaw, K. 1989. Demarginalizing the intersection of race and sex: A Black feminist critique of antidiscrimination doctrine, feminist theory and antiracist politics. *University of Chicago Legal Forum* 40(1):139-167.

Crenshaw, K., N. Gotanda, G. Peller, and K. Thomas (eds.). 1995. *Critical race theory: The key writings that formed the movement.* New York: The New Press.

Creswell, J. W. 2013. Steps in conducting a scholarly mixed methods study. *DBER Speaker Series*, 48.

Cuddy, A. J. C., S. T. Fiske, and P. Glick. 2004. When professionals become mothers, warmth doesn't cut the ice. *Journal of Social Issues* 60(4):701-718.

Delgado, R. 1988. Critical legal studies and the realities of race: Does the fundamental contradiction have a corollary. *Harvard Civil Rights-Civil Liberties Law Review*, 23,407.

Derks, B., N. Ellemers, C. van Laar, and K. de Groot. 2011a. Do sexist organizational cultures create the Queen Bee? *British Journal of Social Psychology* 50(3):519-535.

Derks, B., C. van Laar, N. Ellemers, and K. de Groot. 2011b. Gender-bias primes elicit queen-bee responses among senior policewomen. *Psychological Science* 22(10):1243-1249.

Dillon, E. C., Jr., J. E. Gilbert, J. F. Jackson, and L. J. Charleston. 2015. The State of African Americans in computer science—The need to increase representation. *Computing Research News*, 21(8):2-6.

Eagly, A. H., and A. Mladinic. 1994. Are people prejudiced against women? Some answers from research on attitudes, gender stereotypes, and judgments of competence. *European Review of Social Psychology* 5(1):1-35.

Fiske, S. T. 1999. (Dis)respecting versus (dis)liking: Status and interdependence predict ambivalent stereotypes of competence and warmth. *Journal of Social Issues* 55(3):473-489.

Foschi, M. 1996. Double standards in the evaluation of men and women. *Social Psychology Quarterly* 59(3):237.

Foschi, M. 2000. Double standards for competence: Theory and research. *Annual Review of Sociology* 26:21.

Glaser, B. G. and Strauss, A. 1967. *The discovery of grounded theory: Strategies for qualitative research.* Chicago, IL: Aldine.

Kachchaf, R., L. Ko, A. Hodari, and M. Ong. 2015. Career–life balance for women of color: Experiences in science and engineering academia. *Journal of Diversity in Higher Education* 8(3):175-191.

Levitt, H. M., S. L. Motulsky, F. J. Wertz, S. L. Morrow, and J. G. Ponterotto. 2017. Recommendations for designing and reviewing qualitative research in psychology: Promoting methodological integrity. *Qualitative Psychology* 4(1):2–22.

Malcom, S. M., P. Q. Hall, and J. W. Brown. 1976. *The double bind: The price of being a minority woman in science.* Publication 76-R-3. Washington, DC: American Association for the Advancement of Science. http://web.mit.edu/cortiz/www/Diversity/1975-DoubleBind.pdf.

NASEM (National Academies of Sciences, Engineering, and Medicine). 2013. *Seeking solutions: Maximizing American talent by advancing women of color in academia.* Washington, DC: The National Academies Press.

NASEM. 2016. *Barriers and opportunities for 2-year and 4-year STEM degrees: Systemic change to support students' diverse pathways.* Washington, DC: The National Academies Press.

NASEM. 2020. *Promising practices for addressing the underrepresentation of women in science, engineering, and medicine: Opening doors.* Washington, DC: The National Academies Press.

NRC (National Research Council). 2010. *Gender differences at critical transitions in the careers of science, engineering, and mathematics faculty.* Washington, DC: The National Academies Press.

NRC. 2011. *Expanding underrepresented minority participation: America's science and technology talent at the crossroads.* Washington, DC: The National Academies Press.

NSF (National Science Foundation). 2013. National Center for Science and Engineering Statistics, Scientists and Engineers Statistical Data System. Washington, DC: National Science Foundation. https://www.nsf.gov/statistics/sestat/.

NSF. 2017. *Women, minorities, and persons with disabilities in science and engineering.* Washington, DC: National Science Foundation. https://ncses.nsf.gov/pubs/nsf21321/.

Ong, M. 2010. *The Mini-Symposium on Women of Color in Science, Technology, Engineering, and Mathematics (STEM): A summary of events, findings, and suggestions.* Cambridge, MA: TERC.

Ong, M., C. Wright, L. Espinosa, and G. Orfield. 2011. Inside the double bind: A synthesis of empirical research on undergraduate and graduate women of color in science, technology, engineering, and mathematics. *Harvard Educational Review* 81(2):172-209.

PCAST (President's Council of Advisors on Science and Technology). 2012. *Engage to excel: Producing one million additional college graduates with degrees in science, technology, engineering, and mathematics.* Report to the President.

Slaton, A. E., and A. L. Pawley. 2018. The power and politics of engineering education research design: Saving the "small N." *Engineering Studies* 10(2-3):133-157.

Tulshyan, R., and J. Burey. 2021. Stop telling women they have imposter syndrome. *Harvard Business Review,* https://hbr.org/2021/02/stop-telling-women-they-have-imposter-syndrome.

Williams, J. C. 2014. Hacking tech's diversity problem. *Harvard Business Review* 92(10):94.

Williams, J., and R. Dempsey. 2014. *What works for women at work: Four patterns working women need to know.* New York: New York University Press.

Williams, J., K. W. Phillips, and E. V. Hall. 2014. *Double jeopardy? Gender bias against women of color in science.* San Francisco, CA: Hastings College of Law, University of California.

Winston, C. E., and M. R. Winston. 2012. Cultural psychology of racial ideology in historical perspective: An analytic approach to understanding racialized societies and their psychological effects on lives. In J. Valsiner (Ed.), *Oxford Handbook of Culture and Psychology* (pp. 559-581). New York: Oxford University Press.

Winston, J. A. 1991. Mirror, mirror on the wall: Title VII, Section 1981, and the Intersection of Race and Gender in the Civil Rights Act of 1990. *California Law Review* 79(3):775-805.

2

Literature Review of Research on Girls and Women of Color in Computing, Science, and Technology

This chapter highlights research literature on girls and women of color in technology and computing fields from the last 15 years (2005-2020) in light of the structural and social factors at the K-12, higher education, and workplace levels that hinder or support them. The term *structural factors* refers to institutional or cultural aspects of a given context, including "demographic composition" (Ahuja, 2002; Armstrong et al., 2018) and "basic elements of norms, beliefs, and values that regulate social action" and their impacts (Bernardi et al., 2007, p. 163; see also Parsons, 1951). Structural factors constitute a set of normative and cultural models that define actors' expectations about behaviors when interacting with one another. *Social factors* refer to social and cultural views, experiences, and biases that incorporate and affect external views of girls and women of color (e.g., family support, stereotyping) that are held in society in general, as well as the internal view that girls and women of color have of themselves (e.g., self-efficacy, self-expectations) (Ahuja, 2002; Ragins and Sundstrom, 1989) (Figure 2-1). Research demonstrates that both structural and social factors can affect individuals' decisions and opportunities to enter, persist, and advance in education and careers in CS/tech (Ahuja, 2002; Armstrong et al., 2018).

This chapter is based on findings from multiple projects. First and foremost, it draws from literature identified and summarized by a team of researchers, led by Maria Ong,[1] at TERC[2] in their three-year National Science Foundation–funded project, "Literature Analysis and Synthesis of Women of Color in

---

[1] Ong is a member of the National Academies Committee on Addressing the Underrepresentation of Women of Color in Tech.

[2] TERC is an independent research-based non-profit organization focused on STEM education pre-K-12, postsecondary, and adult education.

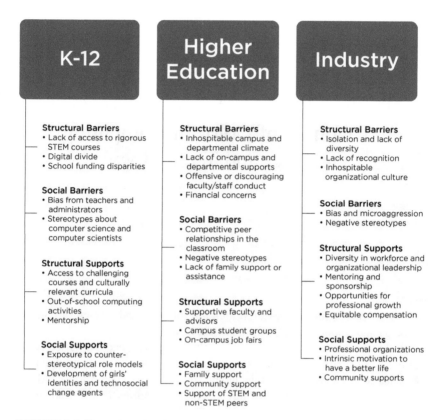

FIGURE 2-1 Examples of social and structural factors that hinder or support women of color in tech.

Technology and Computing."[3] Literature included books, book chapters, peer-reviewed articles, and gray literature.[4] The chapter also draws from a literature review commissioned by the National Academies, written by Heather Lavender,

---

[3] The Literature Analysis and Synthesis of Women of Color in Technology and Computing (NSF Award ID HRD-1760845) team consists of principal investigator Maria Ong, co-principal investigator Nuria Jaumot-Pascual, Audrey Martínez-Gudapakkam, and Christina B. Silva. A detailed description of the methods—including pre-search activities (e.g., testing and selecting electronic literature databases, selecting search terms); conducting literature searches; literature selection (i.e., comparing the content of each literature piece against selection criteria and deciding if it should be discarded or kept for the synthesis); memo writing to summarize each piece for later analysis; codebook development; and coding and thematic analysis—may be found in Ong et al. (2020).

[4] "Gray literature" refers to pieces of literature that are unpublished or published in non-commercial form, such as conference proceedings, dissertations, and reports. They can be of high quality and reflect up-to-date research on understudied topics (Mahood, Van Eerd, and Irvin, 2014), such as women of color in tech. The team's quality appraisal filtering criteria ensured that all studies met high standards for empirical research.

with support from Nuria Jaumot-Pascual and the Literature Analysis and Synthesis of Women of Color in Technology and Computing project team. Other sources include a 2018 data brief, "Women and Girls in Computing" (McAlear et al., 2018), released by the Kapor Center and the Arizona State University Center for Gender Equity in Science and Technology; works arising from the Women of Color in Computing Research Collaborative; and articles shared by members of the National Academies Committee on Addressing the Underrepresentation of Women of Color in Tech. Findings are not exhaustive; they are meant, rather, to give a sense of the general findings about women of color in tech and to inspire a research agenda based on information that is scarce or missing.

The majority of findings reported in this chapter are based on projects that use qualitative research methods with relatively small sample sizes compared to typical projects that use quantitative research methods. The purpose of qualitative research is to gain an in-depth understanding of a phenomenon—often centered around answering *why* and *how* the phenomenon is experienced and the meanings it holds for people—which can be done with small numbers of participants (Creswell, 2013; Glaser and Strauss, 1967). Research based on qualitative methods, such as interview work, aims to understand phenomena by creating categories from the data and then analyzing relationships between categories, while paying attention to the lived experiences of the research participants (Charmaz, 1990, 2006). Qualitative research is a valuable tool for learning about the experiences of women of color in tech, because the understanding of these fields, like many science, technology, engineering, and mathematics (STEM) fields, is hampered by the low numbers of women of color, especially when considering specific institutions, geographic regions, subfields, or races/ethnicities. Qualitative research methods provide an opportunity to examine and understand how the experiences of different groups of women of color vary.

Alice Pawley of Purdue University discussed the value of "learning from small numbers" in STEM (2013, p. 2; also 2019, 2020; Slaton and Pawley, 2018) in a keynote address at one of the project's workshops. She described how vast resources have been poured into large-scale, quantitative STEM studies and interventions, which were largely ineffective for three reasons: studies often used statistical methods of generalization to explain the experiences of underrepresented groups, even when the presence of members of those groups were too low to be statistically significant; interventions emerging from quantitative studies were often attempts to fix the individual instead of the institution; and most studies were based at one type of institution, predominantly white institutions, and did not take into consideration underrepresented students' experiences in other contexts. As an alternative way of handling the small number issue, Pawley urged the use of qualitative methods to understand the ways in which the institutional structure of engineering education and other STEM contexts might be comprehended, illuminated, and changed through a limited set of personal narratives (Pawley, 2013, p. 16).

This chapter takes up that call in both its structure and its findings. At each stage of the computing pipeline—K-12, higher education, and the workplace—institutional, or structural, barriers are prioritized by the committee over social barriers; likewise, structural supports are prioritized over social supports. While the charge of this study and subsequent chapters of the report do not focus on the K-12 stage, the committee believes it is critically important to articulate the supports and barriers that exist for women of color in CS in early educational opportunities and experiences that, later, either support or hinder their further participation in computing in higher education and the workforce. Where literature specific to women of color in CS/tech is scarce, the committee draws on findings in the literature regarding students of color, or women of color, in STEM. The chapter concludes with a brief discussion of non-traditional pathways into tech positions and a list of recommendations arising from the research for practitioners, administrators, and CS/tech education researchers.

K-12 EDUCATION

Nearly a century of research on education indicates that inequality is pervasive throughout the United States' K-12 education system, and Black, Latinx, and Native American students and low-income students receive systemically unequal opportunities for a rigorous and effective education. These barriers are both structural and tied to individual and social factors. While much of this literature captures the specific structural barriers facing Black, Latinx, and Indigenous women in education broadly (and/or how they differ from their male counterparts), it is important to note that women of color face very different educational opportunities in K-12 than their white counterparts. On average, the foundational educational experiences for girls of color from underrepresented groups are significantly different from those of their white peers, providing vastly different opportunities for academic success, early exposure to tech-focused activities, and for entering the tech pipeline. Further, in the current context of the COVID-19 pandemic, there is evidence that existing opportunity gaps for students of color are at risk of widening due to the pervasive digital divide (Common Sense, 2020), the discontinuation of computer science courses (Martin et al., 2020), and the learning loss and poor quality of online learning which are projected to be greatest among Black, Latinx, and Indigenous students (McKinsey, 2020). More research is needed to specifically understand these differences in foundational educational opportunity affecting girls of color.

Structural Barriers for Girls of Color in K-12 Education

Despite efforts to desegregate public schools after the Brown v. Board of Education (1954) decision, schools remain stubbornly segregated by race and income—and unequal (Orfield, 2001). Non-white school districts receive significantly less funding than white school districts, based on the reliance on local property taxes to fund schools. Estimates indicate that this funding gap between non-white school districts and white school districts is as large as $23 billion, with districts spending an average of $2,200 less per student in non-white districts (Baker, 2014; EdBuild, 2019; Kozol, 1992; Morgan and Amerikaner, 2018; NCES, 2012). School funding inequalities are mirrored in the quality and expertise of teachers. There are about twice as many uncredentialed and inexperienced teachers in school districts that serve the highest proportions of low-income and minority students compared with districts with the lowest proportions (Adamson and Darling-Hammond, 2012; Goldhaber, Lavery, and Theobold, 2015). The digital divide is also a pervasive barrier and is pronounced among Black, Latinx, and Native American households, which are significantly less likely to have access to broadband internet and technology devices needed for in-school and out-of-school learning (Common Sense, 2020; PRC, 2012, 2015, 2019). In light of the COVID-19 pandemic, the shift to online learning, and the ways that different groups of girls may be impacted differently (i.e., because of family, community, and geographic characteristics), this foundational structural barrier has become more pronounced and the need to address it more urgent.

Disparities also exist in access to rigorous STEM courses, which vary dramatically by the demographics of schools. Students of color are significantly less likely to have access to advanced placement (AP) and international baccalaureate courses, or to a full range of STEM courses (e.g., physics, calculus) than their peers (ECS, 2017; OCR, 2018). In computer science specifically, students of color, low-income students, and rural students are significantly less likely to have access to computer science courses in their schools, and more specifically lack access to AP computer science courses, which play a critical role in driving interest and preparation in computing in college and career (Code.org 2020; Google/Gallup, 2020; Martin et al., 2015; Scott et al., 2019).

Social Barriers for Girls of Color in K-12 Education

Beyond policies, resources, and course offerings, education research indicates that Black, Latinx, and Indigenous students in K-12 public schools face social barriers stemming from teachers' and administrators' negative biases, belief systems, stereotypes, and expectations. Teachers consistently rate the mathematical proficiency of girls significantly lower than boys (Cimpian et al., 2016) and of girls of color lower than white students—both boys and girls (Copur-Gencturk et al., 2019). Beliefs about the ability of Black, Latinx, and Indigenous students lead counselors and teachers to track students into lower level courses and pathways (Oakes, 1985). Research also indicates that teachers hold lower expectations of college success for Black students than white students (Papageorge et al., 2020). These beliefs can be self-fulfilling and hinder actual student achievement, students' pursuit of STEM fields (Scott and Martin, 2014), and students' rate of degree completion (Gershenson and Papageorge, 2018). Stereotypes and biases also impact how Black girls in particular experience school discipline, with Black girls having suspension rates that are six times higher than white girls (OCR, 2014).

Stereotypes about computer science and computer scientists have been well documented in American society, with assumptions that computer scientists are male, tech oriented, and socially awkward (Cheryan et al., 2013), and that computer science requires brilliance (Leslie et al., 2015) and involves solo, non-collaborative, and non-communal work (Diekman et al., 2010; Margolis and Fisher, 2002). Research shows that classroom cultures with stereotypical cues about computer science have a negative impact on female students' sense of belonging in computer science and their interest in pursuing computing majors and careers (Cheryan et al., 2009; Master et al., 2016). While the research on gendered stereotypes in computing shows that they are present for female students broadly, additional research is needed to fully explore stereotypes and perceptions of computing fields as they relate to girls of color in particular.

As a result of cumulative structural and social barriers, data indicate that girls of color are significantly less likely to take the computer science courses in high school that are significant predictors of entering college and career pathways in this field (Mattern et al., 2011). Girls of color make up approximately 21 percent of the K-12 student population but just 7 percent of all students taking advanced placement computer science courses at the high school level (College Board, 2019; McAlear et al., 2018). Just 77 Native American girls, 3,477 Black girls, and 8,183 Latinx girls took an Advanced Placement Computer Science course in 2019. These numbers have remained static over the period from 2016 through 2021.

Structural Supports

Research demonstrates that students who take the AP Computer Science Principles exam are up to three times more likely to major in computer science and that taking this exam is also a significant predictor of taking the more advanced AP Computer Science A course—a correlation that is particularly strong for Black, Latinx students and girls (Wyatt et al., 2020). Further, students who take the AP Computer Science A exam are seven to eight times more likely to major in computer science in college than their peers who did not take the AP Computer Science A exam (Mattern et al., 2011), indicating that exposure to rigorous computing content in high school is a strong predictor of entering into computing.

Out-of-school exposure to computing activities is also demonstrated to have a positive impact on the interest, aspirations, and knowledge of girls of color. Scott and White (2013) found that culturally responsive after-school programming stimulated the interests and motivations of girls of color to persist in computing—specifically, to learn to master technology and foster an innovative mindset and to disprove negative racial and gender stereotypes about ability in the field of computer science (Scott and White, 2013). Scott and colleagues (2017) found that among girls of color participating in a summer STEM program, levels of computer science interest were low at the outset but increased significantly after multiple exposures to computer science course interventions, although gender differences remained (Scott et al., 2017). Madrigal and colleagues found that an after-school and summer program for Black girls, which included wrap-around mentorship and culturally relevant curricula, yielded promising results for Black girls' interest and confidence in computer science (Madrigal et al., 2020).

Social Supports

A growing body of literature building from the culturally relevant and responsive theoretical frameworks of Gay (2010) and Ladson-Billings (1995) posits that culturally relevant computer science education, curriculum, and pedagogy can improve the classroom experiences, identities, and outcomes of students of color (Scott et al., 2014). Additionally, Ashcraft and colleagues (2017) demonstrated that promising practices for engaging girls of color in computer science were to intentionally develop girls' identities as technosocial change agents and help them understand how technology can be used to advance social justice. There is also some evidence that exposure to counter-stereotypical role models can increase self-concept, attitudes, and career aspirations for women and girls in computer science, although this research is not specific to girls of color (Olsson and Martiny, 2018; Stout et al., 2010). Additional research in both areas will contribute significantly to our understanding of effective interventions for girls of color in K-12 computer science education.

HIGHER EDUCATION

Most research on women of color in technology and computing fields focuses on higher education. Women of color make up 39 percent of the female-identified population in the United States, yet account for less than 10 percent of bachelor's degrees earned in computing and less than 5 percent of doctorates in computing (McAlear et al., 2018). This section first discusses structural and social barriers, followed by structural and social supports. It must be noted that within higher education, most studies are at the undergraduate level; thus that emphasis is reflected here. While research on women of color in technology and computing fields is needed at all levels, it is especially needed at the graduate level.

Structural Barriers for Women of Color in Higher Education

Structural and institutional barriers identified in the literature for women of color at the higher education level include campus and departmental climates experienced as unwelcoming, a scarcity of on-campus and departmental supports aimed specifically at advancing women of color, offensive or discouraging faculty and staff conduct, and the costs related to enrollment in higher education. These are described in detail the sections that follow.

Chilly Campus and Departmental Climates

Several research studies described how chilly, and even hostile, campus and departmental climates contribute to negative experiences of women of color in technology and computing fields in higher education. Institutions and departments that are experienced as unwelcoming often lead students who are women of color to feel excluded and alienated (Ashford, 2016; Charleston et al., 2014b; Thomas, 2016). For example, all 15 African American women in Charleston and colleagues' study agreed that the computer science culture in their respective departments during graduate school at predominantly white institutions were not very welcoming to women, and even less so to African American women. Other studies showed that women of color felt isolated within their departments due to their being the only one, or one of a few, of their gender and/or race or ethnicity (Agbenyega, 2018; Lyon, 2013; Rodriguez, 2015). These women also experienced a sense of lack of commonality with white male peers and cultural disidentification with other students in computer science and other tech departments (Herling, 2011; Tari and Annabi, 2018; Thomas, 2016).

Dearth of On-Campus and Departmental Supports

Another factor contributing to negative experiences of women of color in higher education is the dearth of on-campus formal supports that are prepared to understand and address their unique experiences. Mónica, a first-generation Latina undergraduate in Lyon's (2013) study, lacked information about selecting a major once she arrived at college. Mónica reached out to advising groups at her university, but described her experience as, "I feel like I haven't had a clear advisor. I go to people and I tell them how I feel. . . . But they don't really know what to say back to me" (Lyon, 2013, p. 81). Despite supports at Mónica's university such as assistance in selecting courses and majors as well as an office designed to assist first-generation college students who are the children of migrant workers, she still felt a lack of support given that these services did not address her individual needs, such as helping her choose a major based on her interests. Lyon (2013) also described Kelsey, a first-generation Filipino student interested in informatics, who had access to on-campus support for persistence in college but no guidance on selecting or navigating through a major.

Institutionally sanctioned organizations for students, such as student support programs, are intended to provide students with academic and social support and professional development; for students from underrepresented groups, they may also serve as counterspaces, or safe spaces for belonging (Ong et al., 2018). Some organizations focus on aspects of STEM and gender, such as chapters of the Society of Women Engineers, or STEM and race/ethnicity, such as chapters of the American Indian Science and Engineering Society; however, few are prepared to fully meet the needs of women of color (Herling, 2011). For example, Anu, a Bengali American undergraduate in Ratnabalasuriar's (2012) study, attempted to join the student organization Supporting Women in CS [Computer Science], but she felt out of place and unwelcomed. She stated that "there were very few women in the group. Of the women that were there, most of them were grad students. Almost all of the officers in the organization were men. The students in computer science just aren't very friendly" (p. 127).

Offensive or Discouraging Faculty Conduct

Research showed that faculty sometimes contributed to the negative departmental atmosphere experienced by students who are women of color. Multiple studies found that women of color students in technology and computing fields reported receiving verbal insults or harassment or being treated as invisible by their professors (Ashford, 2016; Charleston et al., 2014a; Hodari et al., 2014). Another way faculty contributed to a negative atmosphere was to engage in institutional microaggressions, such as being aware of inequities against women of color but not taking action to rectify them (Charleston et al., 2014a, 2014b). Faculty's racial and gender biases and attendant lack of support contributed to the perception held by women of color that their departments were hostile (Ashford, 2016; Charleston et al., 2014a, 2014b). For example, women of color reported that professors refused to recommend them for industry positions, regarded teaching women as a chore in comparison to doing their scholarly research, and stated that African American women, in particular, lacked talent and were not intelligent enough to be in computer science (Charleston et al., 2014a; Herling, 2011; Ratnabalasuriar, 2012; Thomas, 2016).

Financial Concerns

Paying for higher education is a key concern for many students and their families. Lack of access to adequate financial resources can serve as a hindrance, especially for underrepresented students of color in STEM, and availability of financial aid for students in need varies across schools (Fenske et al., 2000; Palmer et al., 2011). Students of color from underrepresented groups are more likely to come from families with fewer financial resources, which increases their reliance on paid work while in school, decreases the time they can dedicate to their studies or to participating in STEM organizations and internships, and contributes to stress (AIP, 2020; Estrada et al., 2016; Perna, 2009). Perna (2009), who studied Black women pursuing STEM degrees at Spelman, a historically Black women's college, noted that non-traditional students such as those who commuted, those who were financially independent from their parents, and transfer students were especially vulnerable to financial challenges. In Agbenyega's study (2018), one Latina relayed her financial hardship by providing examples of her difficulties purchasing books for her courses and food insecurity. She stated, "If it wasn't for the pretzel guy on campus, I don't know how I would've lived. That's how I got lunch every day, especially when I didn't have any money" (pp. 161-162). In Foster's (2016) study, two Native women described how they had to pay close attention to the courses they enrolled in at their two-year college to ensure the credits' transferability; they needed to avoid spending more time and money than necessary in obtaining their degrees.

Sufficient financial aid has been highly correlated with persistence of members of underrepresented groups on STEM trajectories (Estrada et al., 2016; Fenske, Porter, and DuBrock, 2000; St. John et al., 2005). Surprisingly, research on financial aid specifically regarding women of color students in technology and computing fields is scarce. Two notable exceptions, studies by Foster (2016) and Lyons (2013), described how communities came together to provide financial support (among other forms of support) to signal their encouragement of women of color entering computing paths. The dearth of published research on the experiences with financial aid of women of color in tech and the role of institutions and communities in addressing women of color's financial need constitute gaps in the literature that need to be addressed.

Social Barriers for Women of Color in Higher Education

Research on women of color in technology and computing fields in higher education reveals a number of social barriers, including challenging relationships with majority peers, challenges related to navigating negative stereotypes, and limitations of family, such as family members' lack of knowledge about the college application process. These factors are discussed below.

Competitive Peer Relationships in the Classroom

In computer science programs an unwelcoming or hostile departmental climate often pervades the classroom environment. Research shows that instead of taking a creative, collaborative approach with learning, peers of women of color students often compete with them, comparing grades rather than discussing course content (Tari and Annabi, 2018), questioning the women's merits because of their ethnicity and gender (Hodari et al., 2016), or questioning their intelligence and abilities (Rodriguez, 2015; Thomas, 2016). Some research reported incidents of peers not wanting to work with or even talk to women of color (Ratnabalasuriar, 2012).

In some cases, because of their interactions with peers, women of color suffered from imposter syndrome, the sense of not legitimately belonging in their field (Ashford, 2016). For instance, a participant in Tari and Annabi's study (2018), whose peers were predominantly white and male, stated that she felt perceived by her peers as a token. Sensing her academic legitimacy was threatened, she stated, "I don't speak out as much as I would have when I went to high school" (p. 4). Tari and Annabi concluded that this participant's legitimacy threat was a major factor in her feeling excluded.

Negative Stereotypes

Studies further reveal how certain gendered and racial stereotypes negatively affected women of color students in technology and computing fields, and how the women responded in order to belong. For example, women who presented themselves in a feminine way, such as wearing high-heeled shoes or dresses, were frequently viewed and treated by peers as less intelligent (Thomas, 2016; Varma et al., 2006). All Latinx women participants in Rodriguez's study (2015) felt there was no middle ground in how their male peers perceived them: The women were seen as either brilliant or stupid, and their competence was called into question even when their work was of the same or higher quality as that of the men. Participants in Lyon's study (2013) combatted the stereotype that the only people who were truly interested in technology and computing fields were male, white and Asian introverted gamers. In response to this stereotype, one Latina student in the study repositioned herself to spend time with classmates who were men instead of the women. Other studies showed that, in the classroom environment, women of color often combatted the perpetuation of the stereotypes of the "angry Black woman" and the "affirmative action" candidate, which contributed to their feelings of not belonging (Ashford, 2016; Charleston et al., 2014a). In response to feelings of exclusion and isolation resulting from falling victim to stereotypes, some women of color adjusted their behavior, language, or attire to be more masculine in order to better fit into their department's environment (Herling, 2011; Thomas, 2016) or considered leaving their program (Rodriguez, 2015).

Lack of Family Support or Assistance

While family can have a very positive effect on women of color who engage in and persist in technology and computing fields (see "Family Supports" below), the research literature suggests that not all family influences are positive. Two studies found that women of color experienced active resistance from family members, such as fathers and stepfathers, who were opposed to the idea of having a female going into a field that they considered inappropriate for females (Agbenyega, 2018; Lyon, 2013). This was the case for a Latina undergraduate in computer engineering in Agbenyega's study, whose father discouraged her from pursuing her major because he saw it as a "men's field." In other cases, family was not an active barrier, but it sometimes lacked the cultural capital of understanding how the U.S. college system works. Lyon (2013) described two women of color who were first-generation college students. They were interested in majoring in informatics but had received little or no information while growing up about the college selection process or resources for finding out such information; the advice these women received from their families was to attend an Ivy League school or community college, with no mention of other viable options, such as state institutions.

Structural Supports for Women of Color in Higher Education

The literature suggests that several structural supports exist to advance women of color in higher education. These include strong, supportive faculty and advisors; STEM and non-STEM campus student groups; and the positive atmosphere and constructive teaching and mentoring provided to students attending historically Black colleges and universities (HBCUs). These factors are described in detail below.

Supportive Faculty and Advisors

While several studies found faculty to be barriers in the education of women of color in technology and computing fields (see "Faculty Conduct" section, above), a few studies identified ways in which faculty members and advisors were very supportive. Ashford (2016) reported women of color participants who learned that their abilities and potential were recognized when they were asked to work in faculty members' labs; these women thus gained valuable advisors. Faculty at HBCUs are known for creating positive learning atmospheres for their students, providing them with professional development and supporting their advancement in STEM, and tech fields such as computer science are no exception (Kvasny et al., 2009; Murray-Thomas, 2018; Wilson, 2016). Finally, a participant in Ratnabalasuriar's study (2012) recalled her computer science department recruiting two women instructors, with mixed results. While these instructors demonstrated a concern for retaining students in the major, the presence of the women instructors did not outweigh the negative climate perceived by the participants.

STEM and Non-STEM Campus Student Groups

Campus student groups, even ones that are not necessarily STEM related, can serve as counterspaces—or safe havens—for women of color and students from other underrepresented groups who may not feel an automatic sense of belonging in their own STEM departments. These groups can help students engage in academic or cultural aspects of campus life in positive ways, and they can help to counter isolation by providing places in which students' experiences are validated and where they feel a sense of safety and belonging (Ong et al., 2018; Solórzano et al., 2000).

Two studies in the literature on technology and computing fields specifically spoke to the importance of campus groups. In the study by Herling (2011), one participant, Gracia, a Hispanic doctoral student, started her own organization at her university for Latinx women in computing. She explained her reasoning: "That was my supportive group, which helped me get through the Ph.D. because we all shared and talked about it" (p. 58). Ninety percent of Herling's participants were members of the group started by Gracia; they attributed their persistence, at least in part, to the group. In Lyon's study (2013), a Latina first-generation college student joined a sorority for Latinas because these peers provided familiarity and comfort to her. The student said of her sorority, "I feel like when I'm tired of being over here in the science field with other people, I feel like I can go 'home' to someone who understands me. . . . I feel like they're supportive to me. And so, they're there for me" (Lyon, 2013, p. 86). More research needs to be done on the benefits of STEM and non-STEM campus student groups for women of color in technology and computing fields overall, and specifically for Asian American, Black and African American, and Native American students.

Historically Black Colleges and Universities

HBCUs make up about 3 percent of all of the colleges and universities in the United States, yet they graduate 25 percent of Black and African American students who receive bachelor's degrees in science and engineering (NSF, 2019; UNCF, n.d.). Similarly, HBCUs graduate a disproportionately high number of Black and African American students who advance to graduate programs in STEM. In 2011, 24 percent of all Black students finishing a doctorate in science and engineering had received their bachelor's degree from an HBCU (Fiegener and Proudfoot, 2013; UNCF, n.d.).

A few studies focused on the experiences of women of color provide insights into reasons for HBCUs' success. In one study, women of color majoring in technology fields at an HBCU reported that their institutions provided an encouraging environment; examples included the teaching of life skills, providing on-campus job fairs, and interacting with faculty who exhibited interest in them (Murray-Thomas, 2018). For example, one participant, Mary, said that a faculty member's encouragement was responsible for her graduating with a combined bachelor's and master's degree in computer science in five years. Mary recalled how the faculty member motivated, supported, and made the students feel at home: "I will not trade having gone to another HBCU for anything. I just feel every time I go back for homecoming; it's like a family" (Murray-Thomas, 2018, p. 66). HBCUs have also provided students with a sense of empowerment and security (Kvasny et al., 2009; Murray-Thomas, 2018; Wilson, 2016). Megan, an application analyst in Kvasny and others' study (2009), stated, "I said [to myself], 'Well, I realize I don't know a lot about the African American culture except for my growing up. I want to know more about the African American culture; I want to be in a little bit more relaxed environment at least for four more years until I have to get out into the real world and deal with the discrimination'" (p.15). And, the reputation of selective HBCUs can aid some women of color in securing jobs. A participant in Middleton's study (2015) perceived that graduation from a top-ranking HBCU aided in her in securing employment within the information technology (IT) company where she worked. More research is needed on the experiences of women of color in technology and computing fields at HBCUs and at other minority-serving institutions,[5] at both the undergraduate and graduate levels.

Social Supports for Women of Color in Higher Education

Research literature on women of color in technology and computing fields in higher education suggests that social supports for women of color include their families, non-STEM peers, other aspects of community, and themselves.

Family Supports

Early exposure to technology and computing through family members had a mostly positive effect for women of color (Agbenyega, 2018; Ashford, 2016; Lyon, 2013; Middleton, 2015; Murray-Thomas, 2018; Thomas, 2016; Thomas et al., 2018). In some cases, family members worked in technology and computing-related careers. For example, one African American woman in Middleton's study (2015) shared that her interest and career in the IT field was sparked by witnessing, as a child, her grandfather work in his television repair shop and seeing what was inside the equipment. Several African American participants in Thomas's study (2016) described their fathers as being a catalyst for their interest in computers and mathematics; one talked about how she "learned to create playlists and clean viruses off of her home computer with her father as her guide" (p. 66). Other studies described how families influenced women of color to pursue computing by emphasizing early on the value of mathematics. For example, one participant in Agbenyega's study (2018), Rosa, a first-generation college student majoring in computer science and daughter of Dominican immigrants, described that her family was overall "mathematically inclined," which helped her strive to excel in mathematics throughout her years in school (p. 161).

Positive family influences were not necessarily specific to technology and computing fields or even STEM. In some cases, the fact that families emphasized the general value of education and of maintaining high academic expectations

---

[5] In this report minority-serving institution refers to historically Black colleges and universities, Hispanic-serving institutions, tribal colleges and universities, and Asian American and Pacific Islander–serving institutions, collectively.

influenced women of color to pursue and persist in technology and computing (Ashford, 2016; Lyon, 2013; Thomas, 2016). Lexi, a participant in Thomas's study (2016), explained how her mother did not have the chance to attend college, but made sure that Lexi pursued college, whatever the subject. In other cases, families, and particularly mothers, served as positive role models for pursuing the educational path that a daughter desired (Foster, 2016; Lyon, 2013; Skervin, 2015). In one notable example in Lyon's study (2013), a participant was inspired to persist in her computer science education by the example of her mother, who, despite protests from her spouse, returned to school to obtain her GED high school equivalency diploma, and then completed a cosmetology certificate.

Non-STEM Peers

To persist in their tech fields, women of color students often draw support for their emotional well-being through a community of non-STEM friends. In several studies women of color reported leaning most consistently on other women (Lyon, 2013; Thomas, 2016). Rankin and Thomas (2020) described several Black women participants in their study who leveraged the social capital of friends outside of their computer science program to meet other Black women on campus. As mentioned above, some women of color joined race- or ethnicity-specific sororities in order to find a sense of "home" on their campus (Lyon, 2013). These findings resonate with other research in STEM education (e.g., Ong et al., 2018; Tate and Linn, 2005) that addressed the ways in which women of color kept their STEM peers separate from the other friends with whom they socialized. The above examples also bring into high relief the importance of counterspaces (safe spaces) for women of color on the campus, even if these individuals or groups are outside of STEM.

Community

The community at large—people who are not family members or non-STEM peers—plays a significant role in the support of women of color in tech. Women of color received support from their community through words of encouragement (Agbenyega, 2018) and financial assistance (Foster, 2016; Lyon, 2013). The three Native women featured in Foster's study (2016) received emotional and financial support from their home communities when they left their reservations to pursue higher education. This support was significant for them because it was emotionally difficult to leave their reservations. One participant, Jaemie, said that support came from elders who told her that she needed to "go get an education, be successful, and come back" (Foster, 2016, p. 120). The women in Foster's study returned to the community as a means of restoring their balance when they had encountered hardships at school. In Lyon's study (2013), Mónica, a first-generation Latina student, received encouragement from her entire community when she learned that she had received a scholarship to a prestigious university. She recounted that she "got support from everyone in the community. Even the Spanish stations were saying my name" (p. 85). This type of support and community recognition was important because Mónica knew her community's sentiment was that computer science was "very difficult and they would be very proud because it's pretty hard to get someone to do something big from our community" (Lyon, 2013, p. 85).

Self

Numerous studies focused on individual or "self" aspects of women of color when reporting on reasons for positive outcomes and supports. The factors include intrinsic qualities and self-efficacy, a sense of creativity and fun and a love for problem solving, science identity development, and the desire to give back.

Intrinsic qualities and self-efficacy. The research literature suggests intrinsic qualities as a key factor in the persistence and success of women of color in tech. Among the intrinsic qualities most frequently cited were those related to motivation, such as being hard working, driven, ambitious, and loving to learn (Foster, 2016; Lyon, 2013; Middleton, 2015; Murray-Thomas, 2018; Thomas, 2016), and those that defined abilities such as being smart or good at math (Agbenyega, 2018; Ashford, 2016; Herling, 2011; Lyon, 2013; Ratnabalasuriar, 2012; Rodriguez, 2015; Zarrett et al., 2006). One study by Charleston and colleagues (2014b) noted that all 15 African American women in their study attributed their own resilience, inner strength, and ability to resist negativity as factors for their endurance in their graduate programs in computing.

Early literature on women of color persisting in STEM focused heavily on aspects of self-efficacy (see Ong et al. (2011) for an overview). In more recent literature on women of color in tech, self-efficacy is identified as a factor in a few studies on women of color, but overall, it is not as heavily emphasized as it once was. For instance, Johnson and colleagues (2008) found that African American women had similar levels of self-efficacy in IT compared to those of African American and white men, contrary to what the researchers had initially hypothesized. Similarly, studies by Zarrett and colleagues (Zarrett and Malanchuck, 2005; Zarrett et al., 2006) and Trauth and colleagues (2012a, 2012b, 2016) found that Black and Hispanic women possessed high confidence and aspirations with regard to computer-related tasks and careers compared to other gender and racial/ethnic groups.

Creativity, fun, and problem solving. Creativity, fun, and problem solving also emerged as interrelated individual factors that influenced the decision of women of color to pursue and persist in technology and computing fields. Researchers found that computer science and coding attracted them because they enjoyed the challenge and the problem-solving opportunities (Agbenyega, 2018; Herling, 2011), the puzzle-like experience (O'Connell, 2018; Skervin, 2015), the logical thinking required (Smith (2016), and the creativity and innovation that is central to the field (Herling, 2011; Smith, 2016). Many of the women of color spoke of how their values of creativity, fun, and problem solving were rooted in their own childhood experiences, such as playing with what were traditionally considered boys' toys like robots and LEGOs (Herling, 2011), tinkering with electronics and computers (Agbenyega, 2018; Herling, 2011; Middleton, 2015; O'Connell, 2018; Ratnabalasuriar, 2012; Thomas et al., 2018), and playing video games (O'Connell, 2018; Ratnabalasuriar, 2012).

Science identity development. Carlone and Johnson's theory of science identity development (2007) may be useful in helping to understand why some women of color persist in technology and computing fields and some do not. In their study of undergraduates who are women of color, they defined three main science identity categories: research scientist, altruistic scientist, and disrupted scientist. Those with research scientist identities recognized themselves as scientists, were focused on the prototypical aspects of science, and were recognized by others as scientists. In contrast, women of color with disrupted scientist identities reported disruptions in their pursuit of a science identity because they were recognized not as scientists, but as representatives of stigmatized groups. Finally, those with altruistic scientist identities recognized themselves as scientists, but they created their own definition of science through the lens of altruistic (philanthropic, self-sacrificing) values.

The research literature on women of color in tech provides a few examples wherein study participants recognized themselves as scientists while they were students (Ashford, 2016; Rodriguez, 2015). Jeanne, a first-generation Haitian American and a postdoctoral researcher at a predominantly white institution, stated that being strong in STEM had always been part of her identity (Ashford, 2016). She shared a story about her father wondering aloud about how she came up with "the most challenging, random [science] questions that would stump the adults" (p. 104). Thus, her science identity was grounded in her intelligence and her competence in STEM. Similarly, Ashley, a Latina undergraduate, used technology-related humor and engaged in computer science–related extra-curricular activities to perform her STEM identity (Rodriguez, 2015). Unfortunately, Ashley simultaneously experienced a disrupted scientist identity, due to the fact that her peers did not recognize her as competent in computer science. In the same study, a participant named Maria described her career identity as aligned with the health care field and her desire to serve her Latinx community, which she identified as a significant part of her overall identity. According to Carlone and Johnson's classification, Maria had an altruistic scientist identity, and this served as a source of motivation for persistence (Carlone and Johnson, 2007).

Desire to give back. According to the research literature, another strong motivation for women of color to pursue and persist in technology and computing fields in higher education is the desire to give back by helping others. This desire to give back connects to Carlone and Johnson's concept of the altruist scientist, whose scientific identity is tied to altruistic values and sees science as a way to improve people's lives. The literature on women of color in tech shows that many women of color see altruistic values as an intrinsic part of their identity as scientists and seek to give back by supporting their communities, mentoring or serving as role models, and supporting those who are like them in some way, such as sharing their same gender and/or race/ethnicity or being interested in similar fields (Agbenyega, 2018; Foster, 2016; Herling, 2011; Hodari et al., 2014, 2015, 2016; Lyon, 2013; Rodriguez, 2015; Skervin, 2015; Thomas, 2016). For example, Abbie, an African American computer science major, emphasized the need to be successful in school in order to be a role model for her younger siblings (Thomas, 2016). Francesca, an Asian American undergraduate student in computer science, used her coding skills to create computer games to aid the learning of middle school algebra students, many of whom were young students of color (Hodari et al., 2015). The research showed that even as women of color engaged in helping others, their giving-back activities contributed equally to their own persistence along their tech trajectory.

WORKPLACE

Studies on women of color in tech at the workplace level are rare and relatively recent; most have been published in the last decade. Similar to the discussion in the section on "Higher Education," findings in this section are presented as structural and social barriers, then structural and social supports.

Structural Barriers for Women of Color in the Workplace

Structural barriers at the workplace level identified in the literature include the lack of diversity and the resulting isolation experienced by women of color, the hypermasculine "bro" culture of tech, and the lack of recognition of the competence and achievements of women of color in these fields.

Lack of Diversity and Resulting Isolation

As described Chapter 1, the technology and computing workforce in the United States is overwhelmingly composed of white men, and, to a lesser degree, Asian or Asian American men (Skervin, 2015). Literature demonstrates that the lack of gender and racial/ethnic diversity in workplaces increases the sense of isolation for women of color in tech and negatively affects their sense of belonging. In O'Connell's study (2018), three women of color "expressed feelings that this intersection of identities enhanced their sense of being the 'other,' and compounded the challenges they face" (p. 76). Women of color felt isolation due to the lack of diversity at their job level. For example, participants in Alegria's (2016) study, who saw few other colleagues who looked like them, stated they felt that women of color were nearly invisible; moreover, those who were there received greater scrutiny in their work than did employees who were white. Participants in Smith's study (2016) reported that they struggled to maintain their identity, or sense of selves, as African American women in their tech workspaces; they described "adjusting their behavior at times to assimilate more into the group, but want[ing] to guard against going too far" (p. 142). Most women of color in Skervin's study (2015), with the exception of Asian women, said they felt tokenized. In other words, they felt "simultaneously a visible representation of their ethnicity, and invisible to their boss for opportunities and promotion" (p. 173).

The isolation of not seeing oneself well represented among peers in the workplace is compounded by lack of diversity in the organization's leadership. In O'Connell's study (2018), Caroline, an East Asian programmer, reflected how persistence in tech was challenging for those like her who embody intersectional, underrepresented identities as women and people of color: "[the isolation is] even more so [as] the level of representation of people like me disappears at higher levels. It's just harder to see yourself there" (p. 72).

"Bro" Culture of Tech

A 2017 study on workers who leave tech found that individuals working in the tech industry experienced and observed more unfairness in the workplace than those employed in non-tech industries, suggesting that tech companies may have significantly more challenges in both culture and employee treatment, and women were much more likely to experience mistreatment of all forms (Scott et al., 2017). The hypermasculine, "bro" culture of technology and computing fields likely contributes to these observations and outcomes. The "bro" culture is well documented, with characteristic behaviors including gender discrimination, sexual harassment, extreme partying, cut-throat competition, bullying, and rule-breaking (Berdahl et al., 2018; Chang, 2019; Payton and Berki, 2019; Sahami, 2018). One widely known documented example of "bro" culture in tech workplace culture is described in a blog post by Susan Fowler, an engineer at Uber, who reported repeated sexual harassment by her manager. Fowler's complaints to the human resources department were not acted upon since the manager was regarded as a top performer at the company (Chang, 2019; Fowler, 2017). Fowler's account, which went viral, put into high relief the institutionalized behaviors that enable "bro" culture to continue. Even with public scrutiny and fall-out (the Uber chief executive officer resigned), in many tech companies, such "bro" behaviors not only are still tolerated, but are normalized and encouraged (Berdahl et al., 2018).

Unsurprisingly, this culture is often blamed for the low representation of women, and especially women of color, in technology and computing fields (Chang, 2019; Smith, 2016; Williams, 2020). In Scott and colleagues' study mentioned above, women of color in particular were much more likely than their peers to be negatively stereotyped, passed over for promotions, and sexually harassed within tech workplaces than their peers (Scott et al., 2017). Skervin's study (2015) showed that women of color participants reported feeling excluded when men would treat them like "the Lady" by noticeably avoiding inappropriate conversations and language in their interactions with them. At the same time, they disliked the "boys' club mentality" where men made women uncomfortable with sexualized conversations. Similarly, Agbenyega (2018) described how women of color participants in her study felt excluded in their workplaces due to their male coworkers' not treating them as "one of the guys" and not inviting them to social-ize outside of work. Such exclusion meant they were left out of opportunities to socially bond with colleagues or discuss work, which, in turn, meant it was more difficult for them to succeed.

Lack of Recognition of Competence and Achievements

A few studies have found that women of color were not given adequate pub-lic recognition by supervisors and leaders for their competence and achievements in technology and computing. Smith's study (2016), for example, found that African American women in IT reported that they felt their professional advance-ment was inhibited by a lack of recognition. While the women received good general work performance reviews, they did not receive public acknowledgment or appreciation for successfully completing specific job projects, as their white male coworkers did; this disparity in public recognition, an example of attribution bias, negatively impacted the women's visibility and advancement (Smith, 2016). O'Connell (2018) reported that women of color, instead of being recognized and rewarded for excelling at soft skills such as communication, teaching, and mentoring in the technology workplace, were punished. The literature also shows instances in which, upon making a mistake, women of color were not provided the second chance that others received (Kvasny et al., 2009) and that supervisors tolerated a lack of productivity from white men but not from women of color (Alegria, 2016; Kvasny et al., 2009; Skervin, 2015).

Social Barriers for Women of Color in the Workplace

Research literature on women of color in technology and computing at the workplace level demonstrates social barriers including biases and microaggres-sions in the workspaces. Specifically, several studies reported that women of color had the credibility of their expertise and competence questioned, had to prove their abilities and worth to peers and supervisors continually, and felt the burden of disproving negative stereotypes of groups to which they belonged.

Biases and Microaggressions

Several research studies describe aspects of hostile work environments for women of color in tech. Sue (2010) defined a hostile work environment against women as a workplace in which "sexual harassment, physical abuse, discriminatory hiring practices" are manifested or where "women [are] subjected to a hostile, predominantly male work environment" (p. 11). Additionally, Sue (2010) argues that as society and employers have become more sensitive to these issues, hostility against women and other marginalized groups has become increasingly expressed in the workplace through microaggressions. Thus, microaggressions— such as messages that people of color are less capable or untrustworthy or that women's contributions are less worthy than men's—must be added to a broadened list of criteria for analysis of hostile work environments.

Three broad, related issues contributing to a hostile work environment for many women of color are described in the sections that follow: the questioning of the credibility of women of color and the constant pressure for them to prove themselves, the pressure on them to disprove stereotypes, and the exclusionary "bro" culture of tech environments.

Credibility and prove-it-again. Questioning the credibility of women of color was particularly pervasive in the literature (McGee, 2017; Middleton, 2015; O'Connell, 2018; Skervin, 2015; Smith, 2016). McGee (2017) identified three women of color tech professionals in her study who had their legitimacy and credibility repeatedly doubted. For example, one Latina participant described how, during her first job out of college, a male coworker pulled her aside and asserted that the only reason she had her job was because of affirmative action. Another participant, an African American woman, likened the recurring invalidation to the movie Ground Hog Day: "There's . . . all the speculation around how you got to be where you are. And that never stops" (pp. 83-84).

In response to having their credibility persistently questioned, many women of color feel pressure in the workplace to continually prove themselves, that is, pressure to provide "more evidence of competence than men in order to be seen as equally competent" (Williams, 2014, p. 5). Williams refers to this particular form of work gender bias as the "prove-it-again" phenomenon (p. 5). Kvasny and colleagues (2009) reported that women of color needed to do more than white male coworkers to be seen as equally qualified. In O'Connell's study (2018), six women of color professionals reported receiving recurrent messages that they were "not technical enough." One African American participant reported how the challenge to prove herself began with the pre-hire interview and continued through her career at her company: "I've been told a lot that I'm not technical enough, which is funny because I write code for a living" (p. 65). A South Asian woman in the same study described the toll that contending with proving-it-again can have on members of underrepresented groups: "People who are minorities in tech feel like they need to prove more of themselves. . . . [T]here's a whole thing at companies that they don't want to lower the bar to hire more diverse candidates. You hear that all the time. So, a part of it also messes with your head in the whole process."

Disproving negative stereotypes. Women of color, especially African American women, are often negatively stereotyped in the tech workplace. According to Williams (2014), women of color in STEM deal with unique, negative stereotypes—for example, that Latinas have fiery tempers, Asian women are meek and mild, and Black women are angry (pp. 7-8). Additionally, all women, including Asian and Asian American women, contend with stereotypes that they are not as intelligent in the workplace as men. For example, Black and Latinx women have the extra burden of disproving the stereotype of laziness (Williams, 2014, pp. 48-49) and broader race-related stereotypes about intellectual ability (Scott et al., 2017).

The literature demonstrates the ways in which women of color in tech assume the burden of addressing stereotypes that result from the embodiment of intersecting identities. Studies have noted how assertiveness is necessary for respect in the tech industry, and is rewarded for people who are not women of color (McGee, 2017; Skervin, 2015), yet others describe the challenge faced by African American women who *are* assertive in the tech workplace and must contend with the stereotype of being seen as an "angry Black woman" or combative when they were simply being assertive in the tech workplace (O'Connell, 2018; Ross, 2014). At the same time, the studies also note that such assertiveness is necessary to be respected in the tech industry, and it is, in fact, rewarded for others (McGee, 2017; Skervin, 2015). To illustrate, one African American participant in Ross's study (2014) described how her colleagues translated her assertiveness as her being an "angry Black woman" who was "trying to prove something or trying to over-exert [her] authority" (p. 113), while the same level of assertiveness from her white female colleague was deemed acceptable by others. Additionally, African American women were burdened with the extra work of ensuring they did not appear angry in their workplace interactions (Ross, 2014).

Structural Supports for Women of Color in the Workplace

The literature discusses four types of structural supports that can help to advance women of color in the workplace: diversity in the workforce and leadership, on-the-job supports of networking and mentoring, opportunities for professional growth, and compensation, described in the sections that follow.

Diversity in the Workforce and Leadership

Research on workplaces suggests that companies that prioritize diversity, equity, and inclusion practices better retain employees from underrepresented groups (Cloutier et al., 2015).[6] Smith (2016), for example, found that African American women expressed that experiencing diversity at all levels in their workplace helped to retain them at their organization and in the field of information technology. Conversely, in O'Connell's study (2018), a lack of diversity in the workplace was seen by women of color as a red flag.

Furthermore, studies have found perceptions by women of color that the leadership of a company influenced their own career path. The race or ethnicity of the company leaders often impacted the perception of women of color in their own treatment. For example, McGee (2017) reported on an Asian American woman who attributed the diversity and culture of the workplace, and her own positive opinion of the company, to the fact that the chief executive officer was Asian. Similarly, Natalie, an African American woman in McGee's study (2017), felt that because her company's chief executive officer was an African American man, she was judged solely on her work performance and not her race. Notably, while Natalie expressed the belief that race was not a factor at her workplace, she observed that women of color experienced a slower rate of promotion (McGee, 2017).

Leaders who are people of color have the potential to adapt non-traditional leadership styles that benefit women of color, among others. For example, an African American woman who was in a leadership position approached giving recognition differently than her peers:

> I make sure that the right parties get credit. And that you will be amply regarded for that idea and for the contributions you make to the team. For example, if you bring in a sale, I make sure that you are amply rewarded for that sale specifically, because that typically causes people to work a little bit harder. So, I am very goal-oriented because of that, and I believe in setting goals for employees and recognizing them when they hit those goals. I think that is a wonderful, wonderful way to motivate folks, because that is what motivates me. I try to give people as much recognition that I possibly can (Ross, 2014, pp. 121-123).

---

[6] We use the New England Resource Center for Higher Education's definitions of diversity, equity, and inclusion. Diversity refers to individual and group/social differences that can be engaged in the service of learning and in the workplace. Equity refers to creating opportunities for equal access and success for historically underrepresented populations through representational equity, resource equity, and equity-mindedness. Inclusion refers to the active, intentional, and ongoing engagement with diversity, which has the potential to increase individuals' awareness, context knowledge, cognitive sophistication, and empathic understanding of the ways individuals interact within systems and institutions (NERCHE, 2016).

This leader was aware of the importance of public recognition for all employees (Carlone and Johnson, 2007) and made sure that she provided this recognition when deserved.

On-the-Job Supports: Networking and Mentoring

Studies of women of color working within the technology field reveal the importance of networking and mentors to both emotionally survive the workplace (Agbenyega, 2018; Foster, 2016; Skervin, 2015) and thrive in their career (Hodari et al., 2014; Middleton, 2015; Smith, 2016). The literature shows that women of color seek to make informed choices in education and career advancement through networking with people in and outside of STEM (Foster, 2016; Middleton, 2015). Participants in Middleton's study (2015) noted that networking was very important, having had a direct role in their obtaining jobs. The need for networking and support was found to be crucial among African American women in Smith's study (2016); they actively sought out informal networks and mentoring relationships within the workplace, often crossing both gender and racial boundaries.

Women of color also actively seek out mentors for leadership training, career advancement, assistance with technical career goals, and interpersonal skills (Hodari et al., 2014; Smith, 2016). For example, in Hodari and colleagues' study (2014), a Latina professional in industry received leadership skills training from an internal mentoring relationship that she proactively formed. Additionally, being mentored earlier in one's career proved to be an effective strategy for navigating the workplace and putting in place a plan for future advancement (Smith, 2016).

Opportunities for Professional Growth

In addition to role models and mentors (Foster, 2016; Hodari et al., 2016; Skervin, 2015; Smith, 2016), career growth and advancement are dependent on several elements of the work environments of women of color, including opportunities for professional development and advanced technical training (Hodari et al., 2016; Skervin, 2015; Smith, 2016). For example, Smith (2016) found that African American women professionals in IT reported that having role models and mentors at a company during the first five years of their career was beneficial to their career growth. After this point, they felt they had outgrown the need for role models or mentors, but they wanted to mentor and serve as role models for more junior African American women in order to increase the numbers of this population in the IT subfields of health care, financial services, education, business consulting, and technology organizations. Smith (2016) found that professional development and advanced technical training helped to retain women of color in the workplace. Because the tech industry is in perpetual transformation, professional development offerings helped to expand their skill set, which then helped lead to career advancement.

Compensation

The median salary in computer and information technology positions in 2019 was $88,240, compared with the median annual wage for all jobs of $39,810 (BLS, 2020). Moreover, jobs in technology and computing are projected to increase 11 percent between 2019 and 2029, which is faster than the average for all occupations, and even faster than most other STEM jobs (BLS, 2020). Unsurprisingly, research studies found that job security and high-earning positions were identified as factors for attracting and retaining women of color in tech, even those graduating with bachelor's degrees (Herling, 2011; Middleton, 2015).

A participant in Thomas's (2016) study, Tamara, learned early in life from her mother how lucrative tech careers could be: "'Oh my God, technology, technology'. Telling me all the jobs go to the technology people in the company and how much money they make . . . and so as a kid I was always into technology because my mom wanted me to learn programs on the computer for that reason" (p. 68). One participant in Middleton's study (2015) witnessed how her peers in high-paying IT jobs had been pulled away from school prior to completing their degrees to pursue these jobs. She remarked, "You know, back then, kids were making a lot money. They weren't even letting people with IT degrees finish college back then, they were like, look . . . we'll start you off with a good salary, just come to work for us" (p. 94).

However, the research also identified a negative side to compensation: while technology and computing provided a "good salary," the salary could also "trap" women of color. Some studies suggested that even as structural barriers prevented women of color from advancing in their companies, they stayed in the tech jobs because they felt they would not have the same quality of life if they left (Middleton, 2015). For example, in Opara et al.'s (2005) study, an African American IT worker at a petrochemical company said, "I can't go any higher in the organization. There is a glass ceiling at this place, but I can't make this same money anywhere else, so I will probably stay where I am" (p. 45). Additionally, a participant of Smith's (2016) study stated

> I feel like computer science is a trap. It is a good field in that it's well paid. But you can burn out after a while so you feel trapped not knowing what else you can do. But what keeps you is that it's the comfort of being well paid for something you are comfortable doing (p. 123).

Opara et al. (2005) found that women of color in IT fields seemed to be satisfied with salary, whereas men were more likely to look at a combination of factors for their job satisfaction.

Social Supports, Personal Drive, and Women
of Color in the Workplace

The literature on women of color in technology and computing in the work-place describes the role of social supports that improve outcomes for women of color such as professional organizations, faith and church, and non-STEM friends, as well as the influence of personal drive to attain a better life. These factors are discussed in the sections that follow.

Professional Organizations

Three studies identified the importance of professional organizations, par-ticularly those that focus on people of color or women in STEM professions, as strong social supports for women of color, particularly Latinx and Black women. Herling (2011) found that the two national organizations, Women in Computer Science and Latinas in Computing, provided Latinx women and Hispanic women professionals with a sense of support and reinforced their computer scientist identities. Similarly, Agbenyega (2018) found that a professional organization serving Hispanic engineers and the National Society of Black Engineers (NSBE) provided Latinx women and Black female professionals in her study with a sense of belonging in STEM, strategies for navigating the STEM workplace, and op-portunities to give back to communities with whom they had something in com-mon. Middleton (2015) also found NSBE played a key role in the persistence of African American women computing professionals in the field. One participant stated, "I think [seeing others who look like me] helped me more than I realized while I was going through it. Because it helped me find people like me, men and women, Black men and women, who were going into this field, where it is predominately dominated by white males" (p. 99).

Faith, Church, and Community

Several studies reported on women of color relying on their spirituality or religious faith to guide their workplace behaviors and as means of coping with barriers and even thriving at the tech workplace (Foster, 2016; Ross, 2014; Skervin, 2015). Some women of color attributed their success to their faith (Ross, 2014; Skervin, 2015). Churches were described as encouraging environments (Murray-Thomas, 2018) and as an empowering source, particularly for African American women (Kvasny et al., 2009). For example, a participant in Ross's (2014) study explained that her faith enabled her "to conquer complicated goals and strive toward better outcomes" (pp. 123-124). African American women in Ross's (2014) study who were in leadership positions at their workplaces used their faith as the guide for treating others kindly and for framing problems in a positive way. One participant reflected how

> In the African American community the church and our religion has been such a large part of our culture, and it is a part of my coping mechanism. What I use to cope is my belief, my faith in God. . . . But not only that, but my faith that tells me that nothing happens by coincidence, and even in the role [as a leader] I am there to make an impartation to the people I interact with. That I am there to have this as a growth and development experience as well (p. 125).

One study by Skervin (2015) illustrated the importance of community composed of family and friends outside of the STEM workforce. These family members and non-STEM friends provided women of color with social supports, a sense of belonging, and spaces to safely discuss their negative experiences in the workplace. Skervin found that the support of a strong community outside of work can help women of color persist and succeed in the workplace. One African American industry professional in the study noted, "[W]hen you have good people around you—outside of the workplace—it really helps you with negativity of the workplace, because you're not just defined by that job" (p. 135).

Desire for a Better Life

While the motivations of women of color for pursuing a career in technology and computing fields vary greatly, one reason recurring in the literature was the desire to have a better life. The meaning of "better life" had several elements, such as wanting a flexible lifestyle that allows a better work-life balance (Ashford, 2016; Thomas, 2016). The most frequently mentioned influences, however, were the related factors of job availability and financial security that technology and computing could offer (Herling, 2011; Kvasny et al., 2009; Middleton, 2015; Murray-Thomas, 2018; Opara et al., 2005; Skervin, 2015; Smith, 2016; Thomas, 2016). For example, Gloria, a Hispanic woman in Herling's study (2011), changed her major to computer science because "being able to find a job after graduation was a huge consideration on my part" (p. 69). In Murray-Thomas's study (2018), Mary, a Black female master's student in computer science, chose her field with the goal of living, with her child, above the poverty line. She stated that "my motivation came from wanting to escape the home environment I grew up in and to have a better life" (p. 87).

NON-TRADITIONAL PATHWAYS TO TECH

Lastly, the committee reviewed the literature on pathways into technology and computing fields that involve job training outside of academic degree programs. Job training currently includes a variety of training initiatives, including, but not limited to, coding bootcamps, such Coding Dojo, which is primarily based on the West Coast, and Boston-based Resilient Coders; web development bootcamps such as Fullstack Academy in New York; apprenticeship programs such Apprenti, which is nationwide; and live-and-learn programs, such as G|Code House in Boston, which is currently under development. Organizations such as Per Scholas, which has multiple training sites throughout the United States, and the Bay Area Video Coalition and Street Tech in California also offer short-term (two- to nine-month) training programs. Programs at community-based and non-profit organizations also offer job training and short IT certificate courses (Chapple, 2006; Kvasny and Chong, 2006). Though some of these initiatives have existed for decades, research literature reporting on them has not kept pace, indicating a gap in the empirical research that needs to be addressed.

The sparse existing research reveals that women of color who took non-traditional pathways into technology and computing often began training in order to escape poverty or to improve their lives (Kvasny and Chong, 2006). Chapple (2006) found while training programs were not exclusive to women of color, they attracted this population due to the promise of employment opportunities enabled by IT skill. However, that promise may or may not be borne out in reality. While Chapple found that most job options immediately available to women of color completing short-term training programs were often low-wage, entry-level IT positions (such as software testing), websites of programs such as Resilient Coders and Apprenti boast an average salary of its participants of around $100,000.

In addition to providing training, non-traditional programs also offer other benefits, such as networking opportunities and professional development. For example, the District of Columbia Web Women, a non-profit organization, caters to women and fosters a space where like-minded women support each other and learn about the rapid changes within technology fields (Sniderman, 2014). Regarding the organization's workshops, Sibyl Edwards, an African American woman study participant, stated, "The workshops help people make themselves more marketable to get hired. And since many of the members are freelance contractors, myself included, we are having more events geared toward contracting, pay, and salary negotiation to empower those groups" (Sniderman, 2014, p. 42).

CONCLUSIONS AND SUMMARY OF NEEDS

The sections that follow provide brief summaries of the literature on girls and women of color in technology and computing fields from the last decade. At each level—K-12, higher education, and the workplace—institutional or structural barriers are prioritized over social barriers; likewise, structural supports are prioritized over social supports. Where literature specific to women of color in tech is sparse, the committee drew on findings in the literature regarding students of color or women of color more generally in STEM. The section concludes with a brief discussion of non-traditional pathways into tech positions and a list of research needs for practitioners, administrators, and technology and computing education researchers.

Summary of K-12 Findings

The research literature shows that structural barriers for girls of color at the K-12 level include de facto segregation in many public school systems, including predominantly non-white schools receiving less funding per pupil and having fewer resources, such as more inexperienced teachers (e.g., EdBuild, 2019) and fewer advanced placement and computer science classes (e.g., Code.org, 2020). Social barriers include negative biases held by teachers about girls of color, which translate into low expectations for STEM achievement and these girls' placement into lower academic tracks (e.g., Copur-Gencturk et al., 2019; Papageorge et al., 2020). Moreover, technology and computing fields have strong stereotypes associated with them—such as its being highly masculine, competitive, non-collaborative, and non-communal (e.g., Master et al., 2016)—that often discourage girls of color from considering the field as a possible career path.

The literature suggested that effective structural supports at the K-12 level for girls of color include course offerings and out-of-school exposure to technology and computing. Research shows that when students, especially girls, Black, and Latinx students, are able to prepare for and take CS AP exams, they are more likely to persist on a technology and computing track and major in computing in college (Mattern et al., 2011). Moreover, studies have found that when girls of color participate in after-school or summer tech programs, they are more likely to express interest in pursuing tech careers (Madrigal et al., 2020; Scott and White, 2013). Social supports include culturally responsive curricula and pedagogy that incorporate into computing tasks the identities of girls of color as technosocial change agents (Ashcraft et al., 2017). Research also suggests that providing students of color with counter-stereotypes improves their self-concept and aspirations in computer science and tech, though more research specifically on girls of color is needed (Olsson and Martiny, 2018).

Summary of Higher Education Findings

Research literature shows that structural barriers for women of color in higher education include experiencing chilly campus and departmental climates (e.g., Ashford, 2016); a dearth of on-campus and departmental supports, such as helpful advisors or student support programs fully prepared to serve them (e.g., Lyon, 2013; Herling, 2011); and hostile or unsupportive faculty (e.g., Charleston et al., 2014a, 2014b). A few studies (e.g., Agbenyega, 2018) hinted at the importance of financial support for women of color in tech, but this aspect of structural support needs to be better studied. Research shows that social barriers for women of color have included negative peer interactions such as having their intelligence questioned (e.g., Rodriguez, 2015), dealing with negative stereotypes (e.g., Thomas, 2016), and not having strong family support (e.g., Agbenyega, 2018).

According to the literature, structural supports for women of color in higher education include supportive faculty and advisors (e.g., Ashford, 2016) and student groups that provided safe places for belonging (e.g., Herling, 2011). A few studies focused specifically on the characteristics and practices of HBCUs that support the success of Black or African American women. HBCUs were shown to provide a supportive environment, a sense of empowerment and security (e.g., Kvasny et al., 2009), and faculty invested in students' success (e.g., Murray-Thomas, 2018). Social supports include family members who supported women of color's STEM interests (e.g., Middleton, 2015) or just cheered them on their technology and computing path (Thomas, 2016); non-STEM peers (e.g., Rankin and Thomas, 2020); and other types of community, such as people in their home-towns (Foster, 2016; Lyon, 2013). Finally, numerous studies focused on qualities of the self, including intrinsic STEM talent and self-efficacy (Ratnabalasuriar, 2012; Zarrett et al., 2006); enjoyment of mathematics, science, and problem solving (e.g., O'Connell, 2018); a strong sense of their scientist identities (e.g., Ashford, 2016); and a desire to give back to their communities (e.g., Hodari et al., 2015).

Summary of Workplace Findings

At the workplace level, research shows that structural barriers for women of color include a lack of diversity and resulting isolation (e.g., Alegria, 2016); the alienating, hypermasculine "bro" culture of tech (Scott et al., 2017); and the lack of recognition of the women's achievements, which translated into slower rates of professional growth and promotions (e.g., Smith, 2016). Social barriers include challenges related to workplace biases and microaggressions, specifically, having their credibility in technology and computing questioned (e.g., McGee, 2017) and needing to disprove negative stereotypes (e.g., Williams, 2014).

The literature suggests that structural supports to advance women of color include diversity in the workforce and leadership (e.g., McGee, 2017); on-the-job supports such as networking and mentoring (e.g., Foster, 2016); opportunities for professional growth like advanced technical training (Smith, 2016); and job security and relative high pay in tech (Middleton, 2015). The literature further revealed that social supports for women of color include professional organiza-tions such as Latinas in Computing or the National Society of Black Engineers (Herling, 2011); faith, church, and community (e.g., Ross, 2014); and a desire for a better life (e.g., Opara et al., 2005). Finally, given the increasing demand for workers with expertise in computing, there is a burgeoning number of short-term, alternative education options arising that enable women of color to enter jobs in the tech sector without having to complete a multiyear academic degree (e.g., Kvasny and Chong, 2006).

Recommendations for Future Research and Funding Needs

The research described in this chapter represents the burgeoning number of studies published in the past decade on girls and women of color in technology and computing fields. However, several gaps in our knowledge base remain on this topic; these gaps are described below. The committee strongly recommends that researchers consider taking up these topics in future research work, and that funders prioritize funding for these topics. Research that helps address the gaps identified by the committee will critically contribute to the knowledge base about how to better support and retain girls and women of color in technology and computing education and careers, and funding for such research should be prioritized.

K-12: Topics for Future Research and Funding

- Differences between girls of color and non-Hispanic white girls with regard to the digital divide, access to computer science courses, and quality of online learning
- Differences between girls of color and non-Hispanic white girls, and between women of color and non-Hispanic white women, with regard to educational experiences during the COVID-19 pandemic
- Intervention components that can positively impact the identity, confidence, interest, and aspirations of girls of color in tech and related fields, including counter-stereotypical role models, culturally relevant computing curricula, access to early childhood education that promotes culturally relevant socioemotional development, and diversity in the computer science teacher workforce

Higher Education: Topics for Future Research and Funding

- Impact of family support/encouragement and other early exposure experiences for women of color in tech and related fields
- How finances and financial aid (e.g., scholarships, loans and debt, salaries) impact women of color's entry into and persistence in tech and related fields
- Experiences of women of color at transition points throughout their academic career (e.g., from K-12 to higher education, from community college to four-year institutions, and from undergraduate to graduate education)
- Experiences of women of color in tech and related fields at the undergraduate and graduate levels at minority-serving institutions
- Experiences of women of color in tech and related fields at technical colleges and community colleges
- Impact of peer mentoring on the success of women of color in tech
- Experiences of graduate students in tech and related disciplines who are women of color
- Experiences of women of color in tech in STEM and non-STEM community and counterspaces (i.e., safe spaces)

Workplace: Topics for Future Research and Funding

- Effective recruitment and hiring of women of color in tech and related fields
- Alternate pathway programs for women of color into tech careers
- Experiences of women of color in tech in STEM and non-STEM community and counterspaces (i.e., safe spaces)
- Award nominations and award receipt rates for women of color in tech
- The intrinsic qualities of women of color that contribute to persistence in tech and related careers
- Women of color in the tech workplace, specifically
  - how they enter the field (having a technical background vs. not having one)
  - promotion rates, experiences with employers, reasons for persistence or attrition
  - how finances (e.g., salaries, pay inequality) impact women of color's entry and persistence in technology and computing

REFERENCES

Adamson, F., and L. Darling-Hammond. 2012. Funding disparities and the inequitable distribution of teachers: Evaluating sources and solutions. *Education Policy Analysis Archives* 20(37):1-41. http://epaa.asu.edu/ojs/article/view/1053.

Agbenyega, E. T. B. 2018. "We are fighters:" Exploring how Latinas use various forms of capital as they strive for success in STEM. PhD dissertation, Temple University.

Ahuja, M. K. 2002. Women in the information technology profession: A literature review, synthesis and research agenda. *European Journal of Information Systems* 11(1):20-34. https://doi.org/10.1057/palgrave.ejis.3000417.

AIP (American Institute of Physics). 2020. *The time is now: Systemic changes to increase African Americans with bachelor's degrees in physics and astronomy.* College Park, MD. https://www.aip.org/sites/default/files/aipcorp/files/teamup-full-report.pdf.

Alegria, S. N. 2016. A mixed methods analysis of the intersections of gender, race, and migration in the high-tech workforce. PhD dissertation. University of Massachusetts, Amherst. https://scholarworks.umass.edu/dissertations_2/614.

Armstrong, D. J., C. K. Riemenschneider, and L. G. Giddens. 2018. The advancement and persistence of women in the information technology profession: An extension of Ahuja's gendered theory of IT career stages. *Information Systems Journal* 28(6):1082-1112. https://doi.org/10.1111/isj.12185.

Ashcraft, C., E. K. Eger, and K. A. Scott. 2017. Becoming technosocial change agents: Intersectionality and culturally responsive pedagogies as vital resources for increasing girls' participation in computing. *Anthropology and Education Quarterly* 48(3):233-251. doi:10.1111/aeq.12197.

Ashford, S. N. 2016. Our counter-life herstories: The experiences of African American women faculty in U.S. computing education. PhD dissertation. University of South Florida. https://scholarcommons.usf.edu/etd/6171/.

Baker, D. 2014. *America's most financially disadvantaged schools and how they got that way.* Washington, DC: Center for American Progress. https://files.eric.ed.gov/fulltext/ED561094.pdf.

Berdahl, J. L., M. Cooper, P. Glick, R. W. Livingston, and J. C. Williams. 2018. Work as a masculinity contest. *Journal of Social Issues* 74(3):422-448. https://doi.org/10.1111/josi.12289.

Bernardi, F., J. J. González, and M. Requena. 2007. The sociology of social structure. In *21st century sociology: A reference handbook*, edited by B. Bryant and D. Peck. Thousand Oaks, CA: Sage. Pp. 162-170.

BLS (Bureau of Labor Statistics). 2020. *Occupational outlook handbook: Computer and information technology occupations.* Washington, DC. https://www.bls.gov/ooh/computer-and-information-technology/home.htm.

Brown v. Board of Education. 347 U.S. 483 (1954). https://scholar.google.com/scholar_case?case=1 2120372216939101759&hl=en&as_sdt=6&as_vis=1&oi=scholarr.

Carlone, H. B., and A. Johnson. 2007. Understanding the science experiences of successful women of color: Science identity as an analytic lens. *Journal of Research in Science Teaching* 44(8):1187-1218. https://doi.org/10.1002/tea.20237.

Chang, E. 2019. *Brotopia: Breaking up the boys' club of Silicon Valley.* New York: Portfolio.

Chapple, K. 2006. Foot in the door, mouse in hand: Low-income women, short-term job training programs, and IT careers. In *Women and information technology: Research on the reasons for under-representation*, edited by W. Aspray and J. Cohoon. Cambridge, MA: MIT Press. Pp. 449-470.

Charleston, L., R. Adserias, N. Lang, and J. Jackson. 2014a. Intersectionality and STEM: The role of race and gender in the academic pursuits of African American women in STEM. *Journal of Progressive Policy and Practice* 2(3):273-293.

Charleston, L. J., P. L. George, J. F. Jackson, J. Berhanu, and M. H. Amechi. 2014b. Navigating un-derrepresented STEM spaces: Experiences of Black women in U.S. computing science higher education programs who actualize success. *Journal of Diversity in Higher Education* 7(3):166-176. https://doi.org/10.1037/a0036632.

Charmaz, K. 1990. "Discovering" chronic illness: Using grounded theory. *Social Science and Medicine* 30:1161-1172. https://doi.org/10.1016/0277-9536(90)90256-R.

Charmaz, K. 2006. *Constructing grounded theory: A practical guide through qualitative analysis.* London, England: Sage.

Cheryan, S., V. C. Plaut, P. G. Davies, and C. M. Steele. 2009. Ambient belonging: How stereotypi-cal cues impact gender participation in computer science. *Journal of Personality and Social Psychology* 97:1045-1060. http://dx.doi.org/10.1037/a0016239.

Cheryan, S., V. C. Plaut, C. Handron, and L. Hudson. 2013. The stereotypical computer scientist: Gendered media representations as a barrier to inclusion for women. *Sex Roles* 69:58-71. http://dx.doi.org/10.1007/s11199-013-0296-x.

Cimpian, J., S. Lubienski, J. Timmer, M. Makowski, and E. Miller. 2016. Have gender gaps in math closed? Achievement, teacher perceptions, and learning behaviors across two ECLS-K cohorts. *AERA Open* October-December 2016, 2(4):1–19. https://doi.org/10.1177/2332858416673617.

Cloutier, O., L. Felusiak, C. Hill, and E. J. Pemberton-Jones. 2015. The importance of devel-oping strategies for employee retention. *Journal of Leadership, Accountability and Ethics* 12(2):119-129.

Code.org. 2020. *2020 State of Computer Science Education: Illuminating Disparities.* Retrieved from https://advocacy.code.org/2020_state_of_cs.pdf

College Board. 2019. *AP Program Participation and Performance Data.* Retrieved from https://research.collegeboard.org/programs/ap/data/archived/ap-2019

Common Sense. 2020. *Closing the Digital Divide in the Age of Distance Learning.* Retrieved from https://www.commonsensemedia.org/sites/default/files/uploads/pdfs/common_sense_media_re-port_final_6_26_7.38am_web_updated.pdf.

Copur-Gencturk, Y., J. Cimpian, and S. Lubienski. 2019. Teachers' bias against the mathematical ability of female, Black and Hispanic students. *Educational Researcher* 49(1):30-43. https://doi.org/10.3102/0013189X19890577.

Creswell, J. W. 2013. *Research design: Qualitative, quantitative, and mixed methods approaches.* Los Angeles, CA: Sage.

Diekman, A. B., E. R. Brown, A. M. Johnston, and E. K. Clark. 2010. Seeking congruity between goals and roles: A new look at why women opt out of science, technology, engineering, and mathematics careers. *Psychological Science* 21:1051-1057. http://dx.doi.org/10.1177/0956797610377342.

ECS (Education Commission of the States). 2017. Percentage of students in high schools which do not offer challenging math and science, by race/ethnicity, 2013-2014. https://vitalsigns.ecs.org/state/united-states/curriculum.

Ed Build. 2019. *$23 Billion*. https://edbuild.org/content/23-billion/full-report.pdf.

Estrada, M., M. Burnett, A. G. Campbell, P. B. Campbell, W. F. Denetclaw, C. G. Gutiérrez, S. Hurtado, G. H. John, J. Matsui, R. McGee, C. M. Okpodu, T. J. Robinson, M. F. Summers, M. Werner-Washburne, and M. Zavala. 2016. Improving underrepresented minority student persistence in STEM. *CBE—Life Sciences Education* 15(3), essay 5. https://doi.org/10.1187/cbe.16-01-0038.

Fenske, R. H., J. D. Porter, and C. P. DuBrock. 2000. Tracking financial aid and persistence of women, minority, and needy students in science, engineering, and mathematics. *Research in Higher Education* 41(1):67-94. https://doi.org/10.1023/A:1007042413040.

Fiegener, M., and S. Proudfoot. 2013. Baccalaureate origins of U.S.-trained science and engineering doctorate recipients. Alexandria, VA: National Center for Science and Engineering Statistics. https://www.nsf.gov/statistics/infbrief/nsf13323/nsf13323.pdf.

Foster, C. 2016. Hybrid spaces for traditional culture and engineering: A narrative exploration of Native American women as agents of change. PhD dissertation. Arizona State University.

Fowler, S. 2017. Reflecting on one very, very strange year at Uber. *Susan Fowler,* February 19. https://www.susanjfowler.com/blog/2017/2/19/reflecting-on-one-very-strange-year-at-uber.

Gay, G. 2010. *Culturally responsive teaching*. 2nd ed. New York: Teachers College Press.

Gershenson, S., and N. Papageorge. 2018. The power of teacher expectations: How racial bias hinders student attainment. *Education Next* 18(1):65-70.

Glaser, B. G., and A. L. Strauss. 1967. *The discovery of grounded theory: Strategies for qualitative research*. New York: Aldine Publishing.

Goldhaber, R., L. Lavery, and R. Theobold. 2015. Uneven playing field? Assessing the teacher quality gap between advantaged and disadvantaged students. *Educational Researcher* 44(5):293-307. https://doi.org/10.3102%2F0013189X15592622.

Google/Gallup. 2020. *Current Perspectives and Continuing Challenges in Computer Science Education in U.S. K-12 Schools*. Retrieved from https://csedu.gallup.com/home.aspx.

Herling, L. 2011. Hispanic women overcoming deterrents to computer science: A phenomenological study. PhD dissertation. University of South Dakota.

Hodari, A. K., M. Ong, L. T. Ko, and R. R. Kachchaf. 2014. New enactments of mentoring and activism: U.S. women of color in computing education and careers. *Proceedings of the Tenth Annual Conference on International Computing Education Research* 1(1):83-90. https://doi.org/10.1145/2632320.2632357.

Hodari, A. K., M. Ong, L. T. Ko, and J. Smith. 2015. Enabling courage: Agentic strategies of women of color in computing. *2015 Research in Equity and Sustained Participation in Engineering, Computing, and Technology (RESPECT)* 1(1):1-7. https://doi.org/10.1109/RESPECT.2015.7296497.

Hodari, A. K., M. Ong, L. T. Ko, and J. M. Smith. 2016. Enacting agency: The strategies of women of color in computing. *Computing in Science and Engineering* 18(3):58-68. https://doi.org/10.1109/MCSE.2016.44.

Johnson, R. D., D. L. Stone, and T. N. Phillips. 2008. Relations among ethnicity, gender, beliefs, attitudes, and intention to pursue a career in information technology. *Journal of Applied Social Psychology* 38(4):999-1022. https://doi.org/10.1111/j.1559-1816.2008.00336.x.

Kozol, J. 1992. *Savage inequalities: Children in America's schools*. New York: Harper Perennial.

Kvasny, L., and J. Chong. 2006. Third world feminist perspectives on information technology. In *Encyclopedia of Gender and Information Technology*. IGI Global. Pp. 1166-1171.

Kvasny, L., E. M. Trauth, and A. J. Morgan. 2009. Power relations in IT education and work: The intersectionality of gender, race, and class. *Journal of Information, Communication and Ethics in Society* 7(2/3):96-118. https://doi.org/10.1108/14779960910955828.

Ladson-Billings, G. 1995. Toward a theory of culturally relevant pedagogy. *American Educational Research Journal* 32(3):465-491. doi:10.3102/00028312032003465.

Leslie, S. J., A. Cimpian, M. Meyer, and E. Freeland. 2015. Expectations of brilliance underlie gender distributions across academic disciplines. *Science* 347:262-265. http://dx.doi.org/10.1126/science.1261375.

Lyon, L. A. 2013. Sociocultural influences on undergraduate women's entry into a computer science major. PhD dissertation. University of Washington.

Madrigal, V., R. Yamaguchi, A. Hall, and J. Burge. 2020. Promoting and supporting computer science among middle school girls of color: Initial findings from BRIGHT-CS. In *Proceedings of the 51st ACM Technical Symposium on Computer Science Education (SIGCSE'20)*. https://doi.org/10.1145/3328778.3366855.

Mahood, Q., D. Van Eerd, and E. Irvin. 2014. Searching for grey literature for systematic reviews: Challenges and benefits. *Research Synthesis Methods* 5(3):221-234. https://doi.org/10.1002/jrsm.1106.

Margolis, J., and A. Fisher. 2002. *Unlocking the clubhouse: Women in computing.* Cambridge, MA: MIT Press.

Martin, A., S. Koshy, L. Hinton, A. Scot, B. Twarek, and K. Davis. 2020. *Teacher Perspectives on COVID-19's Impact on K-12 Computer Science Education.* Kapor Center. https://mk0kaporcenter5ld71a.kinstacdn.com/wp-content/uploads/2020/10/KC20005_csta-report_final.pdf.

Martin, A., F. McAlear, and A. Scott. 2015. Path not found: *Disparities in access to computer science courses in California high schools.* Oakland, CA: Level Playing Field Institute.

Master, A., S. Cheryan, and A. Meltzoff. 2016. Computing whether she belongs: Stereotypes undermine girls' interest and sense of belonging in computer science. *Journal of Educational Psychology* 108(3):424-437.

Mattern, K. D., E. J. Shaw, and M. Ewing. 2011. *Is AP exam participation and performance related to choice of college major?* New York: College Board. https://eric.ed.gov/?id=ED561044.

McAlear, F., A. Scott, K. Scott, and S. Weiss. 2018. *Women and girls of color in computing.* Data brief. Kapor Center. https://www.wocincomputing.org/#data-brief.

McGee, K. 2017. The influence of race, ethnicity, and gender on advancement in information technology. PhD dissertation. Fielding Graduate University.

McKinsey. 2020. *COVID-19 and Learning Loss: Disparities Grow and Students Need Help.* https://www.mckinsey.com/industries/public-and-social-sector/our-insights/covid-19-and-learning-loss-disparities-grow-and-students-need-help.

Middleton, K. L. 2015. Factors of influence for occupational attainment of African-American women in information technology. PhD dissertation. Robert Morris University.

Morgan, I., and A. Amerikaner. 2018. *Funding gaps: An analysis of school funding equity across the U.S. and within each state.* Education Trust. https://edtrustmain.s3.us-east-2.amazonaws.com/wp-content/uploads/2014/09/20180601/Funding-Gaps-2018-Report-UPDATED.pdf.

Murray-Thomas, L. 2018. Jumping over hurdles to get to the finish line: Experiences influencing Black female advanced STEM degree attainment. PhD dissertation. California State University, Long Beach.

NCES (National Center for Education Statistics). 2012. School district current expenditures per pupil with and without adjustments for federal revenues by poverty and race/ethnicity characteristics. Table A-1, B-1. https://nces.ed.gov/edfin/Fy11_12_tables.asp.

NERCHE (New England Resource Center for Higher Education). 2016. NERCHE Self-Assessment for the Institutionalization of Diversity, Equity, and Inclusion in Higher Education. https://www.wpi.edu/sites/default/files/Project_Inclusion_NERCHE_Rubric-Self-Assessment-2016.pdf.

NSF (National Science Foundation). 2019. *Women, minorities, and persons with disabilities in science and engineering.* https://ncses.nsf.gov/pubs/nsf19304/digest.

O'Connell, P. E. 2018. The lived experience of female programmers working in the San Francisco Bay area: An interpretive phenomenological analytic study. PhD dissertation. Wright Institute.

Oakes, J. 1985. *Keeping track: How schools structure inequality.* New Haven, CT: Yale University Press.

OCR (Office of Civil Rights). 2014. Civil rights data collection data snapshot: School discipline. https://ocrdata.ed.gov/assets/downloads/CRDC-School-Discipline-Snapshot.pdf.

OCR. 2018. 2015-16 civil rights data collection: STEM course taking. Figure 4, Percentage of high schools offering mathematics and science courses. Washington, DC: Department of Education. https://www2.ed.gov/about/offices/list/ocr/docs/stem-course-taking.pdf.

Olsson, M., and S. Martiny. 2018. Does exposure to counterstereotypical role models influence girls' and women's gender stereotypes and career choices? A review of social psychological research. *Frontiers in Psychology* 9:2264. https://doi.org/10.3389/fpsyg.2018.02264.

Ong, M., C. Wright, L. Espinosa, and G. Orfield. 2011. Inside the double bind: A synthesis of empirical research on undergraduate and graduate women of color in science, technology, engineering, and mathematics. *Harvard Educational Review* 81(2):172-209. https://doi.org/10.17763/haer.81.2.t022245n7x4752v2.

Ong, M., J. M. Smith, and L. T. Ko. 2018. Counterspaces for women of color in STEM higher education: Marginal and central spaces for persistence and success. *Journal of Research in Science Teaching* 55(2):206-245. https://doi.org/10.1002/tea.21417.

Ong, M., N. Jaumot-Pascual, and L. T. Ko. 2020. Research literature on women of color in undergraduate engineering education: A systematic thematic synthesis. *Journal of Engineering Education* 109(3):347-615. https://doi.org/10.1002/jee.20345.

Opara, E., V. Etnyre, and M. A. Rob. 2005. Career in information systems: An analysis of job satisfaction among African American males and African American females. *Journal of Information Technology Management* 16(1):39-47.

Orfield, G. 2001. *Schools more separate: Consequences of a decade of resegregation.* Cambridge, MA: Harvard University, Civil Rights Project. https://eric.ed.gov/?id=ED459217.

Palmer, R. T., D. C. Maramba, and T. E. Dancy. 2011. A qualitative investigation of factors promoting the retention and persistence of students of color in STEM. *Journal of Negro Education* 80(4):491-504. https://www.jstor.org/stable/41341155.

Papageorge, N., S. Gershenson, and K. Kang. 2020. Teacher expectations matter. *Review of Economics and Statistics* 102(2):234-251.

Parsons, T. 1951. *The Social System.* New York: Free Press.

Pawley, A. L. 2013. Learning from small numbers of underrepresented students' stories: Discussing a method to learn about institutional structure through narrative. Paper presented at the 2013 American Society for Engineering Education Annual Conference and Exposition, Atlanta, GA. https://peer.asee.org/19030.

Pawley, A. L. 2019. Learning from small numbers: Studying ruling relations that gender and race the structure of U.S. engineering education. *Journal of Engineering Education* 108(1):13-31. https://doi.org/10.1002/jee.20247.

Pawley, A. L. 2020. Learning from small numbers: Theory-informed insights on gender and race. Keynote address at National Academies Workshop on Addressing the Underrepresentation of Women of Color in Technology. Washington, DC.

Payton, F. C., and E. Berki. 2019. Countering the negative image of women in computing. *Communications of the ACM* 62(5):56-63. https://doi.org/10.1145/3319422.

Perna, L., V. Lundy-Wagner, N. D. Drezner, M. Gasman, S. Yoon, E. Bose, and S. Gary. 2009. The contribution of HBCUs to the preparation of African American women for STEM careers: A case study. *Research in Higher Education* 50(1):1-23. https://doi.org/10.1007/s11162-008-9110-y.

PRC (Pew Research Center). 2012. *Digital Differences*. Retrieved from https://www.pewresearch.org/internet/2012/04/13/digital-differences/.

PRC. 2015. *Home Broadband, 2015*. Retrieved from https://www.pewresearch.org/internet/2015/12/21/home-broadband-2015/

PRC. 2019. *Mobile Technology and Home Broadband, 2019*. https://www.pewresearch.org/internet/2019/06/13/mobile-technology-and-home-broadband-2019/.

Ragins, B. R., and E. Sundstrom. 1989. Gender and power in organizations: A longitudinal perspective. *Psychological Bulletin* 105(1):51-88.

Rankin, Y. A., and J. O. Thomas. 2020. The intersectional experiences of Black women in computing. *Proceedings of the 51st ACM Technical Symposium on Computer Science Education* 1(1):199-205. https://doi.org/10.1145/3328778.3366873.

Ratnabalasuriar, S. 2012. Forging paths through hostile territory: Intersections of women's identities pursuing post-secondary computing. PhD dissertation. Arizona State University.

Rodriguez, S. 2015. Las mujeres in the STEM pipeline: How Latina college students who persist in STEM majors develop and sustain their science identities. PhD dissertation. University of Texas, Austin.

Ross, T. L. 2014. African American females leveraging adaptive leadership skills in information technology: A qualitative study. PhD dissertation. University of Phoenix.

Sahami, M. 2018. Paving a path to more inclusive computing. *ACM Inroads* 9(4):85-88. https://doi.org/10.1145/3230700.

Scott, A., and A. Martin. 2014. Perceived barriers to higher education in science, technology, engineering, and mathematics. *Journal of Women and Minorities in Science and Engineering* 20(3):235-256.

Scott, A., F. Kapor Klein, and U. Onovakpuri. 2017. *Tech leavers study*. Kapor Center. https://mk0kaporcenter5ld71a.kinstacdn.com/wp-content/uploads/2017/08/TechLeavers2017.pdf.

Scott, A., A. Martin, F. McAlear, and S. Koshy. 2017. Broadening participation in computing: Experiences of girls of color. Proceedings of the Innovation and Technology in Computer Science Education Conference, July 2017. Bologna, Italy. https://mk0kaporcenter5ld71a.kinstacdn.com/wp-content/uploads/2017/12/ITiCSE-Paper-Allison-Scott.pdf.

Scott, A., S. Koshy, M. Rao, L. Hinton, J. Flapan, A. Martin, and F. McAlear. 2019. *Computer Science in California's Schools: An Analysis of Access, Enrollment, and Equity*. Retrieved from https://mk0kaporcenter5ld71a.kinstacdn.com/wp-content/uploads/2019/06/Computer-Science-in-California-Schools.pdf.

Scott, K., and M. White. 2013. COMPUGIRLS' standpoint: Culturally responsive computing and its effect on girls of color. *Urban Education* 48(5):657-681. https://doi.org/10.1177/0042085913491219.

Scott, K., E. Sheridan, and K. Clark. 2014. Culturally responsive computing: A theory revisited. *Learning Media and Technology* 40(4):412-436. https://doi.org/10.1080/17439884.2014.924966.

Skervin, A. E. 2015. Success factors for women of color information technology leaders in corporate America. PhD dissertation. Walden University. https://scholarworks.waldenu.edu/dissertations/365/.

Slaton, A. E., and A. L. Pawley. 2018. The power and politics of engineering education research design: Saving the "small N." *Engineering Studies* 10(2-3):133-157.

Smith, A. D. 2016. Exploring the retention and career persistence factors of African American women in information technology: A multiple case study. PhD dissertation. Ohio State University. http://rave.ohiolink.edu/etdc/view?acc_num=osu1462814233.

Sniderman, D. 2014. DC web women: Helping women who work in tech. *IEEE Women in Engineering Magazine* 8(2):40-43. https://ieeexplore.ieee.org/document/6949714.

Solórzano, D., M. Ceja, and T. Yosso. 2000. Critical race theory, racial microaggressions, and campus racial climate: The experiences of African American college students. *Journal of Negro Education* 69(1):60-73. http://www.jstor.org/stable/2696265.

St. John, E. P., M. B. Paulsen, and D. F. Carter. 2005. Diversity, college costs, and postsecondary opportunity: An examination of the financial nexus between college choice and persistence for African Americans and whites. *Journal of Higher Education* 76(5):545-569. https://doi.org/10.1353/jhe.2005.0035.

Stout, J., N. Dasgupta, M. Hunsinger, and M. McManus. 2010. STEMing the tide: Using ingroup experts to inoculate women's self-concept in science, technology, engineering, and mathematics (STEM). *Journal of Personality and Social Psychology* 100(2):255-270. doi: 10.1037/a0021385.

Sue, D. W. 2010. *Microaggressions in everyday life: Race, gender, and sexual orientation.* Hoboken, NJ: John Wiley and Sons.

Tari, M., and H. Annabi. 2018. Someone on my level: How women of color describe the role of teaching assistants in creating inclusive technology courses. Paper presented at the 24th Americas Conference on Information Systems, New Orleans, LA.

Tate, E. D., and M. Linn. 2005. How does identity shape the experiences of women of color engineering students? *Journal of Science Education and Technology* 14(1):483-493. https://doi.org/10.1007/s10956-005-0223-1.

Thomas, J. O., N. Joseph, A. Williams, and J. Burge. 2018. Speaking truth to power: Exploring the intersectional experiences of Black women in computing. *2018 Research on Equity and Sustained Participation in Engineering, Computing, and Technology (RESPECT)* 1(1):1-8. https://doi.org/10.1109/RESPECT.2018.8491718.

Thomas, S. S. 2016. An examination of the factors that influence African American females to pursue postsecondary and secondary information communications technology education. PhD dissertation. Texas A&M University. http://hdl.handle.net/1969.1/156994.

Trauth, E. M., C. C. Cain, K. D. Joshi, L. Kvasny, and K. Booth. 2012a. Embracing intersectionality in gender and IT career choice research. *Proceedings of the 50th Annual Conference on Computers and People Research* 1(1):199-212. https://doi.org/10.1145/2214091.2214141.

Trauth, E. M., C. Cain, K. D. Joshi, L. Kvasny, and K. Booth. 2012b. Understanding underrepresentation in IT through intersectionality. *Proceedings of the 2012 iConference* 1(1):56-62. https://doi.org/10.1145/2132176.2132184.

Trauth, E. M., C. C. Cain, K. D. Joshi, L. Kvasny, and K. M. Booth. 2016. The influence of gender-ethnic intersectionality on gender stereotypes about IT skills and knowledge. *ACM SIGMIS Database* 47(3):9-39. https://doi.org/10.1145/2980783.2980785.

UNCF (United Negro College Fund). n.d. The numbers don't lie: HBCUs are changing the college landscape. https://uncf.org/the-latest/the-numbers-dont-lie-hbcus-are-changing-the-college-landscape.

Varma, R., A. Prasad, and D. Kapur. 2006. Confronting the "socialization" barrier: Cross-ethnic differences in undergraduate women's preference for IT education. In *Women and information technology: Research on underrepresentation*, edited by J. Cohoon and W. Aspray. Cambridge, MA: MIT Press. Pp. 301-322.

Williams, J. C. 2014. Double jeopardy? An empirical study with implications for the debates over implicit bias and intersectionality. *Harvard Journal of Law and Gender* 37(1):185-242. http://repository.uchastings.edu/faculty_scholarship/1278.

Williams, J. C. 2020. Workplace experiences of women of color in tech. Presentation at the Women of Color in Computing Collaborative. Summit, Women of Color in Computing: Barriers and Solutions in K-12 and Higher Education, September 21, 2020. Virtual Summit.

Wilson, R. S. 2016. Understanding the experiences and perceptions of African American female STEM majors at single-sex HBCU. PhD dissertation. Morgan State University.

Wyatt, J., J. Feng, and M. Ewing. 2020. *AP Computer Science Principles and the STEM and Computer Science Pipelines.* Retrieved from https://apcentral.collegeboard.org/pdf/ap-csp-and-stem-cs-pipelines.pdf?course=ap-csp-and-stem-cs-pipelines.pdf?course=ap-computer-science-principles

Zarrett, N. R., and O. Malanchuk. 2005. Who's computing? Gender and race differences in young adults' decisions to pursue an information technology career. *New Directions for Child and Adolescent Development* 2005(11):65-84. https://doi.org/10.1002/cd.150.

Zarrett, N., O. Malanchuk, P. E. Davis-Kean, and J. Eccles. 2006. Examining the gender gap in IT by race: Young adults' decisions to pursue an IT career. In J. M. Cohoon (ed.), *Women and information technology: Research on underrepresentation.*

3

Challenging Assumptions Around the Recruitment, Retention, and Advancement of Women of Color in Higher Education

Applying an intersectional lens to higher education makes sense of the experiences of women of color by situating them within the confines of place (e.g., department, school, college) and position (e.g., undergraduate, graduate student, faculty member), and by examining how they interact with organizing features such as meetings, laboratories, geographical location, and modes of working in academia. For example, how a Latinx, female computer science faculty member journeys through a predominantly white institution seeking tenure and promotion may yield markedly different experiences than the same individual navigating this experience in a minority-serving institution[1] as a graduate student. The race and ethnic constitution of higher education institutions are only two categories requiring consideration. Variability in levels of research funding, size, departmental structure (interdisciplinary vs. single-discipline enclaves), whether a higher education institution is public or private, and any combination of these variables (e.g., a predominantly white all women's college such as Smith College or a historically Black all women's college like Bennett College) also shapes the experiences of women of color. There is no one-size-fits-all approach to the

---

[1] Minority-serving institution (MSI), refers to (1) historically defined MSIs, those that were established with the expressed purpose of providing access to higher education for a specific racial minority group and (2) enrollment-designated MSIs, those that are federally recognized as MSIs based on student enrollment thresholds. Historically defined MSIs include Historically Black Colleges and Universities (HBCUs) and Tribal Colleges and Universities (TCUs). Enrollment-Designated MSIs include Hispanic-Serving Institutions (HSIs), Alaska Native-Serving and Native Hawaiian-Serving Institutions (ANNHIs), Asian American and Native American Pacific Islander-Serving Institutions (AANAPISIs), Predominantly Black Institutions (PBIs), and Native American-Serving Nontribal Institutions (NASNTIs). In this report discussion of minority-serving institutions is targeted to HBCUs, HSIs, TCUs, and AANAPISIs, collectively.

recruitment, retention, and advancement of women of color in higher education that can be used for all women belonging to any one group. Intragroup difference requires nuanced strategies that recognize explicit and implicit forms of oppression and how these acts may appear differently based on the micro-, meso-, and/or macro-level in which they occur (e.g., at the level of classrooms; departments; and/or universities, states, or geographic regions, respectively).

This chapter discusses three statements identified by the committee that are frequently used by higher education leaders to describe barriers to the recruitment, retention, and advancement of women of color (both students and faculty) in technology and computing fields in their institutions (Box 3-1), and discusses assumptions often cited to underpin these statements. The committee drew on data, both statistical and empirical, to examine assumptions related to challenges that higher education leaders and institutions face in efforts to increase the representation of women of color in higher education computing departments. The committee also reviewed evidence that counters these assumptions and identified promising practices and recommendations for higher education institutions. In cases where there was a dearth of data focused on women of color in tech, the committee relied on data with contextual information drawn from the lived experiences of women of color and on research literature focused on science, technology, engineering, and mathematics (STEM) courses of study and professions more broadly. This chapter offers recommendations for intentional action that leaders in higher education can take to improve the recruitment, retention, and advancement of women students of color pursuing technology and computing degrees and women of color faculty in these fields.

APPLYING AN INTERSECTIONAL LENS TO THE EXPERIENCES OF WOMEN OF COLOR IN HIGHER EDUCATION

As discussed in Chapter 1, the committee used the critical race theory of intersectionality as an analytic framework to interpret evidence about the under-representation of women of color in technology and computing fields. Power is a core construct of intersectionality and provides a lens to analyze the experiences of women of color in higher education (Collins, 2019). Crenshaw (1991) described power as relationships. As individuals interact with each other, they develop interpersonal patterns, and meaning is made along various axes such as race, gender, ethnicity, indigeneity, social class, and other social categories. Research illustrates how, through interactions with classmates and professors, women of color are treated as powerless individuals lacking "testimonial authority," a term used to describe the credibility granted to a person or group by other socially dominant groups (Kidd and Carel, 2017). This phenomenon has been well documented in computer science departments (Whitecraft and Williams, 2010). The resultant power dynamic is most evident during critical transition points—periods during which significant changes in academic and career tra-

BOX 3-1
Three Key Statements Heard in Higher Education to Excuse the Underrepresentation of Women of Color in Tech

1. Assertion: "We cannot find qualified women of color."
   Stated assumptions:
   - Women of color do not pursue tech degrees due to a lack of encouragement or interest.
   - Women of color in tech are more interested in "soft" computing sub-disciplines such as computing education, human factors, and human-computer interaction.
   - Faculty who are women of color are willingly opting out of academic careers in tech.

2. Assertions: "We just hire the best," "we are color blind," and the like.
   Stated assumptions:
   - Recruitment and advancement decisions are based solely on merit.

3. Assertion: "We have made progress."
   Stated assumption:
   - Increases in diversity numbers mean that efforts to recruit, retain, and advance students and faculty who are women of color are working as needed.

These three statements are often heard in higher education to explain the underrepresentation of women of color in tech on campus, but in reality they excuse institutions from assessing the policies and practices that may lead to or perpetuate this underrepresentation. While few women of color pursue degrees and academic careers in tech disciplines, the structural and social factors that can deter them (e.g., admissions criteria, access to internet, insufficient financial aid, hostile culture, etc.) are often overlooked. When they do enter academia, women of color face pressure to take on additional roles outside of the classroom around their minority status (e.g., serving on hiring committees, serving as a representative of their demographic group to speak to potential new hires) which can limit their advancement. For those institutions that do see growth in the number of women of color in tech, the increase in diversity can lead them to believe that their strategies are working; however, measuring diversity can be misleading—for example, lumping together non-U.S.-born students of color with U.S.-born students of color—and does not capture enough information to determine whether institutional efforts to recruit, retain, and advance students and faculty who are women of color have been successful.

jectories can occur in the lives of students and faculty—such as transitions from community college to a four-year college, from undergraduate to graduate or graduate to faculty, and in promotion and tenure.

Power also influences how individuals relate to structures such as institutional hierarchy or tenure and promotion. Analyses of the extent to which individuals follow the policies and procedures that subjugate women of color or push them out of higher education can reveal what undergirds the power structure. Recognizing the disciplinary, cultural, and structural domains by which the policies and procedures operate is also key to an intersectional endeavor and will reveal not only the complexity of contexts but also how pernicious strategies and systems operate at all levels. Enobong (Anna) Branch, senior vice president for equity at Rutgers University who presented to the committee in April 2020, discussed the problematic nature of using the pipeline metaphor to describe the trajectory from K-12 to higher education. The pipeline metaphor frames women of color as passive participants in an unchangeable system rather than individuals who must actively challenge, engage, and interact with existing systems. Branch proposed an alternative framework for articulating the agency of women of color and the challenges they face—a road with exits, pathways, and potholes. This framework recognizes multiple entry and exit points, challenges that may be experienced differently by different individuals, and challenges that are universal. It recognizes that the ability to navigate even universal challenges can be affected by an individual's experience, agency, self-efficacy, and skill sets (Branch, 2020). The committee adopted this perspective in its activities and analysis.

Through a multitude of structures and policies, higher education institutions have been established and maintained endowing certain individuals with power, financial support, and privilege provided they assimilate the cultural mores of the context. Those in power set the rules, standardize regulations, and maintain the status quo. Promotion and tenure committees, institutional review boards, department chairs, deans, and other authoritative leaders maintain a coveted position within this ecosystem. Endowed as gatekeepers, these individuals are privileged—and at most institutions include few women of color. However, power is not only about status or role. While data clearly illustrate how few women of color obtain leadership roles in technology and computing in higher education, the few women of color who do obtain positions of power endure more oppressive environments than their white peers (Dotson, 2011). To improve the recruitment, retention, and advancement of women of color in higher education it is critical to analyze and offer solutions as to how women of color can have space to narrate their experiences and be heard as testimonial authorities.

A discussion of each key statement excusing the underrepresentation of women of color in tech, and its underlying assumption(s), follows.

ASSERTION: "WE CANNOT FIND QUALIFIED WOMEN OF COLOR"

A lack of qualified women of color is often cited as a reason for disproportionately low numbers of both students and faculty members. The committee examines the validity of some of the key assumptions that are used to support this assertion in the sections that follow.

Assumption: Women of Color Don't Pursue Tech Degrees Due to a Lack of Encouragement or Interest

The number of women of color who graduate with computer science degrees is low.[2] In 2018, women of color constituted 18 percent of the overall population in the United States; however, they make up less than 10 percent of all bachelor's degrees earned in computing, with Latinx women the most underrepresented in computing bachelor's degree completion rates relative to their population in postsecondary education. At the doctoral level, women earned 21 percent of all doctorates in computing with less than 5 percent awarded to Black, Latinx, Native American/Alaskan Native, or Native Hawaiian/Pacific Islander women (McAlear et al., 2018). However, students from underrepresented groups aspire to major in STEM in college at the same rates as their white and Asian American peers, and have done so since the late 1980s (NAS, NAE, and IOM, 2011, p. 4), and women of color express strong interest in science and engineering fields and greater intention to major in these fields in postsecondary education than do white women (Malcom and Malcom, 2011, p. 163).

Within higher education, one excuse given for the lack of representation of women of color has to do with the influence of the family—the argument being that young girls of color, as early as middle school, are not encouraged by their families to pursue careers in computing, thus the seed is not planted early enough. Encouragement from family can indeed contribute to a student being interested in a career—it contributes to a sense of self-efficacy (e.g., "I can study this"), and it is important that young women of color know early on that a career in tech is possible (see Chapter 2). It is also true that interest in STEM careers can be expressed as early as middle school; however, this assumption perpetuates the notion that interest begins in middle school and continues in a linear trajectory through high school and college, leading to a career in STEM, and implies that there is a pipeline to a career in STEM that has early leaks (Harackiewicz et al., 2012). This assumption overlooks the structural and social factors that can deter women of color from pursuing computing degrees.

Other factors can also influence whether young girls feel encouraged to pursue tech as a potential career and whether they feel accepted within tech environments. During its February 2020 workshop, the committee heard from Kyla McMullen, assistant professor and director of the SoundPad Lab at the

---

[2] See https://www.nsf.gov/statistics/2017/nsf17310/digest/fod-wmreg/ai-an-women-by-field.cfm.

University of Florida. McMullen presented data from a 2012 study by the Girl Scout Research Institute showing that 62 percent of African American girls felt their teachers were unsupportive of their career interests compared to 73 percent of white girls who felt their teachers were supportive. She also highlighted the role of personal culture, how it may sometimes conflict with the culture of tech and science for classrooms, and how that conflict can potentially discourage girls of color from pursuing technology and computing careers. For example, students whose cultures value cooperation more than competition may not feel that they fit in in fields where competition, individual assertiveness, and highlighting one's own contributions are valued more than collaboration with peers (McMullen, 2020). The lived experiences of these students later in life may have a greater influence on their decision to pursue a computing major than early encouragement.

Challenges to Starting

Much of the literature related to underrepresentation of women in tech and STEM focuses on deficits in the contexts or capabilities of women of color (e.g., programs that are too difficult, lack of encouragement from family, family obligations, lack of parental or other role models, and lack of access) rather than structures and systems that impede progress and result in persisting inequity in the number of women of color in tech majors (Figure 3-1). For example, admissions to academic programs can be biased and present an obstacle to women of color at the very start of their pursuit of a tech degree. Admissions criteria for many higher education institutions often include standardized test scores; however, research has demonstrated that racial and gender bias may show up in standardized tests in a number of ways (e.g., questions that use expressions that are more common in white society or use of a multiple choice format) and can underpredict future academic success at the postsecondary level (NEA, 2021; Sukumaran, 2020). For example, Holloway and colleagues (2014) found that at a large Midwestern public university while student applications to engineering departments increased 46 percent over a five-year period, admission of women increased by only 23 percent. (While the study did include women of color, the results were not disaggregated by race and gender.) They also found that biases that affected admission of women into programs, when addressed, had a positive impact and increased the number of women admitted. Of interest, they reported that the cognitive metrics of women applicants had statistically significant higher values than the men in grade point average, class rank, and Scholastic Aptitude Test (SAT) verbal scores.

Education infrastructure during the K-12 years can influence outcomes for women of color and can also inform an understanding of barriers to tech fields in higher education. Disparities in early educational and life experiences (such as a lack of educational opportunities) of social norming suggesting that girls, particularly girls of color, are not as well suited for technology and computing

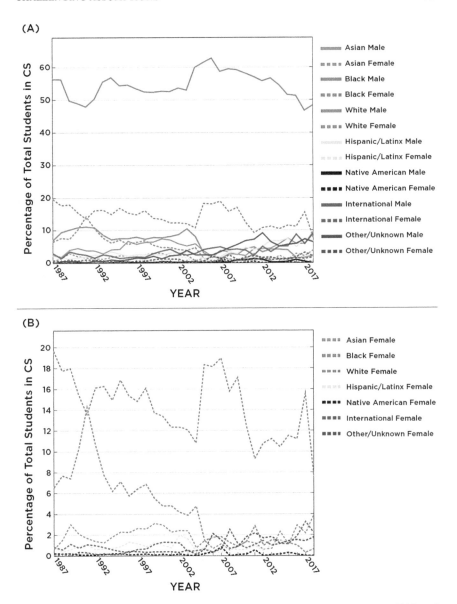

FIGURE 3-1 Annual changes in computer science demographics between 1987 and 2018: MIDFIELD database analysis. (A) Men and women; (B) women only.
SOURCE: Ross (2020).

fields, can shape later life experiences and opportunities for women of color as students and as they pursue postsecondary education or seek to advance into careers as faculty. Kamau Bobb, senior director of Constellations Center for Equity in Computing at Georgia Tech, presented to the committee in May 2020 to discuss the connections between secondary school and higher education and the importance of understanding the contexts of where students in higher education are coming from. Using Atlanta as an example, Bobb highlighted how unequal allocation of educational resources and opportunities among schools even within the same city can disproportionately and negatively impact students of color—for example, when students have less access to advanced placement courses or lack of access to internet and computing devices at home. For students who attend these schools, these inequalities in infrastructure can lead to underpreparation and lack of access to rigorous coursework for students who are motivated and capable, which can have lasting effects that affect outcomes in higher education (Bobb, 2020).

The committee also heard from speakers from its June 2020 workshop who discussed infrastructure challenges experienced by Native students. Twyla Baker, president of Nueta Hidatsa Sahnish College, also cited lack of wireless internet access, and in some cases even cell phone towers, as a major infrastructure challenge for students and their communities and a potential barrier to successful preparation and persistence in higher education (Baker, 2020; Delgado-Olson, 2020). Historically, communities located on tribal lands have had lower levels of access, to telecommunications services relative to other populations (Adcock, 2014). As mentioned in Chapter 2, the lack of or insufficient access to reliable broadband internet service in rural areas can be a significant barrier to learning for Native American students. The digital divide is notable in minority-serving schools, particularly in tribal schools (Varma and Galindo-Sanchez, 2006). In interviews with Native American undergraduate students enrolled in a computing program at six nontribal and tribal universities, lack of exposure to computing and technology was cited by female respondents more frequently than male respondents. Female respondents from tribal sites reported a lack of resources more than respondents from nontribal sites (Varma and Galindo-Sanchez, 2006). Andrea Delgado-Olson, founder and chair of Native American Women in Computing, discussed how the digital divide on reservations impacts students and communities, discussing how the needs of communities may differ since the nature of this challenge can differ by region and within regions.

Challenges at Transition Points

Students from underrepresented groups face obstacles at transition points after entering higher education (Hurtado et al., 2007). A significant number of women of color pursuing tech degrees are non-traditional or "post-traditional" (NASEM, 2020) and do not fit the old notion of full-time enrollment in college, coming straight from high school, living on campus, without an outside job, and graduating in four years. Women of color as well as women in general are more heavily concentrated in community colleges and therefore face transitions that can put up barriers to their progress.

There are numerous factors that may influence women of color to begin higher education in community colleges (such as family obligations, access to support structures, and lack of sufficient financial aid at more expensive schools) even when they could otherwise begin at a four-year institution (Gates, 2020). Women of color represent a disproportionate number of students who received an associate's degree at a community college prior to earning a STEM baccalaureate degree (NASEM, 2016; Reyes, 2011). During the 1990s and early 2000s, part-time enrollment grew for women in community colleges, with particularly strong increases for African American, Hispanic, and Asian American women (Reyes, 2011).

As a result, a significant number of women of color start their academic careers in community colleges and then transfer to four-year institutions. A transfer usually requires declaring a major, which can be influenced by amount of financial aid, number of credits that can be transferred, and even availability of preparatory courses in the community college for the desired major. Thus the transition to four-year colleges can lead to many women of color dropping out or changing majors to fields other than technology and computing. Challenges such as these, rather than lack of interest in computing, create obstacles for many women of color who might otherwise continue the pursuit of a technology and computing degrees or careers.

Challenges During the Undergraduate Years Due to Institutional Culture and Climate

According to one longitudinal study of students enrolled in postsecondary education, 32 percent of students who declared engineering and engineering technology majors and 28 percent of students who declared computer and information sciences majors changed majors within three years of initial enrollment (NCES, 2017, 2020). While the decision to change majors may be influenced by a number of factors, often it is not an issue of academic preparation (e.g., school selectivity or rigor) that prevents women of color from pursuing their interests but rather institutional climates. Institutional climates are a manifestation of a college or university's culture, and are created and sustained by physical structures, policies, underlying values, and social norms. Climates can influence student performance, engagement, and persistence (NASEM, 2016, p. 60). As discussed in Chapter 2, unwelcoming campus and departmental climates can be structural barriers for the success of women of color in technology and computing in higher education. In Cohoon's study (2001) across computer science departments in Virginia, cultural factors had an impact on retaining women students in computing programs at the undergraduate level. The study found that departmental factors that affect gendered attrition include the availability of same-sex peer support; faculty characteristics and behaviors; and institutional and community environment.

Several presentations to the committee during its four public workshops in spring 2020 also highlighted the role of institutional climate and peer attitudes and how they affect the persistence of women of color. For example, Gregory Walton, associate professor of psychology at Stanford University, highlighted how the belief that one does not belong can become self-fulfilling. He also presented examples of statements from Black women in higher education on their experiences in that environment, including: "my input is undervalued because it's not 'mainstream' or 'relatable'" and "I'm more qualified than my peers but I'm treated as less than" (Walton, 2020). Once women of color are in their four-year academic programs, they may face "social pain."[3] In situations where students are from underrepresented groups, their social identities are more salient to both minority and majority group members (Hurtado et al., 1996). Being the "only" (or having "solo status") brings more scrutiny of their performance and might trigger the stereotype threat associated with their group (Hurtado et al., 2007; Steele, 1997; Thompson and Sekaquaptewa, 2002). Such barriers can be more pronounced for women of color who find themselves in a "double bind" of race and gender marginalization when navigating the STEM culture (Malcom et al., 1976; Williams et al., 2014). Other challenges include low self-confidence (or low self-efficacy) and financial aid struggles (Ong et al., 2020; Reyes, 2011).

Assumption: Women of Color in Tech Are More Interested in "Soft" Computing Sub-Disciplines

Women and people from underrepresented groups are more highly represented in applied computing areas,[4] such as computing education, human factors, and human-computer interaction, fields often characterized as "soft" sub-disciplines that may not be valued as highly in the field. Publication patterns show that women are more likely to publish in conferences on human factors, seen as a less prestigious field and considered in alignment with feminine stereotypes, while men are more likely to publish in conferences on algorithms, a field with higher prestige and considered in alignment with masculine stereotypes (Cohoon et al., 2011). These trends lead some observers to conclude that the smaller presence of women overall and women of color in particular in less prestigious technology and computing fields is due simply to their interests and preferences.

---

[3] Ong and colleagues (2020) defined this concept in four categories: being the only one, being made invisible, stereotype threat and being spotlighted, and discrimination and harassment.

[4] See the Computing Research Association's Taulbee Survey for more data on enrollment and degree completion in information, computer science and computer engineering and in providing salary and demographic data for faculty. Statistics given include gender and ethnicity breakdowns. https://cra.org/resources/taulbee-survey/ (accessed October 7, 2021).

Computing education and human factors are examples of popular computing sub-disciplines where women publish. Two conferences of the Association for Computing Machinery's Special Interest Group on Computer Science Education, the Technical Symposium on Computer Science Education and the Conference on Innovation and Technology in Computer Science Education, have high percentages of women authors at 29 percent and 32 percent, respectively. Women typically outnumber men among teachers, and publications in computing education also have high participation from women. Human factors has strong ties to psychology, another area where women outnumber men. Differences in authorship along gender lines are also often present in the popular press, helping cement the notion that women are not interested in technical areas of technology and computing fields. For example, an article in *Forbes* emphasized the notion that women can be in tech without being an engineer or developer (McCullough, 2016). While statements like these encourage women and young girls to consider STEM careers, they reinforce the artificial division within computer sciences that more technical, and possibly higher-paying jobs, belong to men, and women should focus on the more social, applied, feminine jobs.

However, the committee did not attribute the greater participation of women of color in specific areas of computing solely to their interests. Rather, women may lean toward human-related computing areas because these areas afford fulfillment of communal goals or because they are where they see role models, receive mentoring, or simply feel welcomed and thus face fewer social pains (Corbett and Hill, 2015; Ong et al., 2020). These sub-disciplines in computing have been better at mentoring and supporting women. Evidence suggests that a few women of color pioneers in some computing sub-disciplines—possibly brought into computing through interdisciplinary research—have found a home and played a significant role as sponsors to bring in others like them into computing. As these areas grew, the numbers grew and the spaces became more welcoming. However, women may suffer during tenure process as a result of leaning into "soft" computing sub-disciplines. For example, a core technical research grant and an education grant are seen in a very different light, the latter often seen as a second class citizen. Information Systems/Information Sciences is another area which is often seen as a second class citizen in computing. This could be due to campus cultures that elevate some departments over others.

The power of role models also helps explain increased numbers of women of color in some areas of computing. For example, the Computer-Human Interaction Mentoring Workshop has brought together a talented group of students from underrepresented groups doing research in human-computer interaction, beginning in 2010 and 2012 and continuing in 2018 and 2020.[5] Many participants have noted the impact that these workshops had in their careers, highlighting the networking, social connection, and support derived from each other. Most

---

[5] See: https://chime2020.wordpress.com/.

of these students are now in tenure-track positions in academia or in industry research positions (Metoyer et al., 2019).

More study is needed to determine why these sub-disciplines in computing have been better at mentoring and supporting women, to understand whether strategies that attract women of color to these sub-disciplines could be applied to other technical sub-disciplines where women of color are less represented, and to understand the role that women of color play in these spaces.

Lack of Recognition of the Relevance of Communal Goals in Computing Sub-Disciplines

A report by Corbett and Hill (2015) highlights findings from a number of researchers that show that work with a clear social purpose that prioritizes communal goals over career goals and that allows one to help and work with others is preferred more by women than by men. Although there are opportunities for computing jobs to provide these kinds of opportunities to fulfill communal goals, these jobs are typically not viewed that way. More often jobs in computing and other tech fields are thought of as careers with limited opportunities to work with others and contribute to one's community. While this may be true in some cases, this perception may help to partly explain the continued underrepresentation of women in these fields (Corbett and Hill, 2015). Data presented to the committee also support the idea that a lack of persistence in tech can be motivated by culture. Kathy DeerInWater presented preliminary data from research that found that Native American women and two-spirit individuals were motivated to remain in computing when their work allowed them to give back to the community—an important component of Native identities. These findings support results from previous work in this area (DeerInWater, 2020). Research found that many women of color place great value on connecting work to community impact, activism, and outreach, especially in their home communities and as a tool for improving conditions in the workplace for other women of color (Gates, 2020; Ong et al., 2020).

The communal goal congruity perspective suggests that an important aspect of the decision of whether or not to pursue a career in a STEM field is the belief that STEM careers do not fulfill communal goals. To the extent that individuals anticipate and experience STEM fields as fulfilling their valued communal goals, they will be more likely to enter and persist in such fields (Diekman et al., 2017). Drawing on the literature related to women in STEM, evidence suggests that because women may place a greater value on communal goals than men, women may be more likely to decide against careers in STEM in favor of careers in other fields (Corbett and Hill, 2015). This may hold true for women of color, in particular, as well. Thus, the reason for women's relatively greater increased interest in computing sub-disciplines such as computing education, human factors, and human-computer interaction may be because of their perception that these areas will allow them to meet highly valued goals.

A study examining interest in computing among middle school African American girls found that students who saw the application of computing as relevant to their lives showed increased interest in applied computing areas like human-factors and human-computer interaction. The study included creating a hand-sketched user interface for an iPhone app that included text chat, video chat, and/or audio chat features (Robinson, 2015; Robinson and Pérez-Quiñones, 2014). Participants showed high levels of ownership of their designs, creating names for the app (e.g., Chirper, Musicana, Mingle) and images that reflected their own personality (Figure 3-2). Before this activity, the girls expressed the opinion that computer science was boring. Once they were able to "visualize their designs as realistic applications" and interact with the study director, a young African American woman from the same region as the participants, their perception of computing changed. In a survey at the end of the study, they expressed that computing was "cool" and were willing to explore it further. The research demonstrates how exposure to an applied area of computing can change the perception of the discipline and build interest (Robinson, 2015).

Assumption: Women of Color Are Willingly Opting Out of Academic Careers in Tech

Although the proportion of science degrees granted to women has increased, there is a persistent disparity between the number of women receiving doctoral

FIGURE 3-2 Samples of user interface prototypes designed by African American middle school girls.
SOURCE: Robinson (2015).

degrees and those hired as junior faculty. This enduring gap suggests that the problem of underrepresentation of women of color among tenured STEM faculty will not resolve itself solely by more generations of women moving on an upward trajectory through academia; rather, it suggests that women's advancement within academic science may be actively impeded (Moss-Racusin et al., 2012). There is an assumption that any time women of color in tech leave higher education careers or choose not to pursue faculty or leadership positions, they are willingly opting out. The illusion of choice as the sole motivating force persists as a result of the failure to recognize how existing systems as well as organizational and academic climates that serve to limit opportunities for women of color prevent their entry, create unwelcoming environments, and effectively force them out.

It is often assumed that women of color who leave their careers in academia early (e.g., before making full professor) do so because they prioritize attending to family needs (such as taking care of kids or elderly family members) over their career aspirations. While work-family factors have been found to play an important role in women's exits from engineering (Hunt, 2010), family factors do not account for the *majority* of exits from STEM jobs (Ashcraft et al., 2016; Glass et al., 2013). Rather, there are a multitude of structural and systemic factors and conditions affecting these decisions (e.g., organizational culture, workplace bias, lack of advancement). The prevalence of this assumption leads to reform efforts focusing on the individual rather than the structures the individual must endure. Focusing on individual choice at the exclusion of systemic challenges maintains important blind spots. The following sections elaborate on some of these blind spots regarding institutional structures, such as gatekeeping, workplace hostility, lack of access to power, and institutional policies that may negatively impact women of color.

Gatekeeping

Gatekeeping, a commonplace process used by academic programs to establish and adhere to protocols and policies designed to evaluate students' suitability for a professional practice (Hunt and Nicodemus, 2014), starts early and continues along individuals' career trajectories. However, it can take the form of academic elitism in the hiring process for faculty whereby the academic system devalues people whose doctoral degrees are not from prestigious schools—a group disproportionately including women and underrepresented minorities—thus limiting these individuals' opportunities (Clauset et al., 2015). Higher education is a primary site of "opportunity hoarding" where members of the dominant group serve as gatekeepers for access to resources (Riegle-Crumb et al., 2019). When faculty and administrators serve as gatekeepers, their behaviors may serve to discourage and disadvantage women of color (Canning et al., 2019). Women faculty of color can be subjected to epistemic gatekeeping when their scholarship and contributions are minimized and devalued (Dotson, 2012, 2014).

Organizational and Systemic Structures Affecting Academic Climates

Earlier in this chapter institutional climates are discussed that cause social pain that deter women of color from *pursuing* degrees in technology and computing fields, but once women of color are on a pathway to enter academia or are faculty or administrators in these fields, they also often contend with unsupportive and hostile climates that can push them to leave their career trajectories in tech fields. A number of large-scale empirical research studies have documented the persistent and pervasive problem of hostile culture. For example, at least half of all women academics in STEM (versus 19 percent of men academics in STEM) report experiencing sexual harassment, and even greater numbers (78 percent) of women in male-dominated STEM workplaces report experiencing gender-based discrimination (Funk and Parker, 2018; NASEM, 2018a). Also discussed above, several studies have investigated stereotypes encountered by women of color. Black women, for example, are reported as being seen as defiant, angry, or incompetent, and Asian American women are often perceived as not being natural leaders or as being eternal foreigners. Awareness of these stereotypes causes many women to change their behaviors in the workplace to avoid these types of characterizations. Nevertheless, both student and faculty women of color express feelings of isolation and tokenism in the classroom and in the workplace (Ong et al., 2020).

Women of color are often placed in disadvantaged positions due to historical and current discrimination and a lack of access to power and positions of leadership (McGee, 2020, 2021). A number of experts invited to present to the committee at its public workshops highlighted the importance of access to power within academia and the role that access to decision-making power can play in improving outcomes for women of color. Stephanie Adams, dean of the School of Engineering and Computer Science at the University of Texas at Dallas, discussed the importance of accountability in higher education. Leadership at the institutional and departmental level plays a role in shaping the diversity of higher education institutions, cultivating a climate of inclusivity, and creating policies that encourage and reward practices that increase equity (Adams, 2020; Gilbert, 2020; Su et al., 2015). Consistency of leadership in academia, from the chancellor/president down through the provost and deans to department chairs, is important. Constant changes in leadership can lead to changes in strategic plans, departmental foci, degree programs, and requirements, which make it increasingly difficult for students and faculty who are already at risk for retention to persist.

Women of color are underrepresented in these positions of leadership. While not specific to higher education, data show that women of color report slower or even non-existent promotion rates relative to their peers and are more likely to have their credibility questioned by their peers and leaders. Different groups of women of color may experience different challenges related to accessing power or moving into positions of leadership (ACE, 2021; Hill et al., 2016). Asian American women, for example, while overrepresented at the bachelor's level and at the career entry level, face significant challenges advancing to positions of leadership and fare worse in advancement than other groups of women of color (Kim et al., 2018; Peck, 2020). Although a growing number of women have begun to occupy top-level leadership positions, research suggests that these women are likely to be promoted into positions of leadership where the risk of failure is high or in times of crisis—a phenomenon referred to as the glass cliff (Ryan et al., 2016). Women of color who want to advance may end up taking on unwinnable roles without sufficient institutional supports in place.

ASSERTIONS: "WE JUST HIRE THE BEST" OR "WE ARE COLOR BLIND"

An assertion of equality and fairness underpins many of the assumptions surrounding hiring, tenure, and promotion policies and practices. The committee examines these assumptions in the sections that follow.

Assumption: Recruitment and Advancement Decisions Are Based on Meritocracy

The ideology of meritocracy has embedded within it two tacit assumptions: (1) equal opportunity for all, and (2) success is the result of hard work and intelligence. These assumptions lead to unintended outcomes, both of which warrant further exploration. For women, there is a significant threat to academic and career progression when meritocratic explanations are substituted for structural explanations of inequality (Cech and Blair-Loy, 2010). Ample evidence supports the notion that "structural racism in STEM often manifests as meritocracy" (McGee, 2020, p. 635). The illusion of meritocracy ignores patterns of inequality throughout higher education and society, and how structural inequality rewards and privileges those who have wealth and access to influential connections.

Challenging meritocracy is most often met defensively and seen as an attack on the concept of merit. When the challengers are women of color, they often face hostility and backlash for confrontation and, in fact, for their very presence. However, the assumption that decisions are based on meritocracy leads to numerous non-meritocratic outcomes for women of color in STEM: unequal opportunities for entry and advancement, exclusionary and unsupportive cultures, unequal distribution of rewards when deserved, and situations where organizations that purport themselves to be meritocratic, but are not, reinforce members' beliefs that they and their organizations are fair and objective (Kang and Kaplan, 2019). Holding on to the assumption of meritocracy also prevents the pursuit of meaningful systemic change. Not surprisingly, those who benefit most from the assumption of meritocracy are least likely to challenge it—for to do so would require admitting that their own merit might be in question. In challenging meritocracy, the claims of merit-based advancement must be disrupted along with the presumption of equal opportunities, equal access to connections, and equitable distribution of rewards.

Inequitable Burden of Service and Mentoring Activities

Because of their scarcity, women of color in faculty positions can face demanding out-of-class instructional loads. Unlike white male faculty members, many women of color are called on to advise students of color and others studying in similar fields and to handle minority and gender affairs. Institutional reward systems often preference faculty who spend more time on research and scholarship in tenure decisions over faculty who spend more time on service that is assigned to meet institutional needs (Jaschick, 2010; Turner, 2002, p. 82). Women of color are often asked to take on a large number of service commitments (e.g., serving on hiring committees, serving as a representative of both their gender and their race/ethnicity to speak to potential new hires, serving as an advisor to student groups focused on their identity) which take up time that faculty without such obligations can use for research and publishing. The small numbers of women faculty of color compel them to serve simultaneously as a role model for their profession, race, and gender. In many instances, extra service, diversity work, and higher teaching and mentoring loads are less likely to lead to tenure or to prestigious positions related to committee service, and serve to limit career progression and success (NRC, 2013). Junior faculty members are particularly at risk.

Tenure and Promotion Policies

Institutional policies may also negatively affect whether women of color are granted tenure. So-called stop-the-clock policies have been used as a way to support both men and women faculty who are new parents by providing extra time before a decision is made whether to grant them tenure. However, taking personal or family leave time or needing to shift attention to caregiving responsibilities (e.g., for children or elderly family members) can lead to women being viewed as less committed to their careers (Rosser, 2004). Consequently, many women on the tenure track opt out of using policies meant to counteract negative effects on their career progression such as stop-the-clock policies (Wasburn, 2004).

Moreover, research has shown that men may actually benefit more from these policies as they are more likely to utilize the extra time to publish more (potentially raising the bar for others) rather than for its intended purpose (Jaschik, 2016). In addition, research suggests that women are not exiting these careers primarily for family concerns, but that when they do leave due to family obligations they might have made different choices if more flexible options to support these competing responsibilities had been available (Ashcraft et al., 2016). In theory, this is a policy that could also be used to grant extra time to account for relatively heavy faculty service obligations. However, more research is needed to determine whether uptake of this policy in broader contexts would lead to increases in the number of women of color being granted tenure.

Another area that warrants further research is whether bias and inequitable treatment in the evaluation of faculty scholarship affect the progression of women of color in academia. Settles et al. (2020) finds that scholarly devaluation[6] is more common to women and faculty of color. In their study of over 100 faculty of color at a research-intensive institution they found that some types of scholarship were evaluated as lower quality because of the topic (e.g., social problem focused), method (e.g., qualitative), publication outlet (e.g., top-tier journals or presses), and whether it was grant funded. The study found that scholarly devaluation occurred through evaluation processes that the university used to make merit, promotion, and tenure decisions (Settles et al., 2020).

ASSERTION: "WE HAVE MADE PROGRESS"

Defining metrics of success and collecting adequate data are key steps to accurately measuring progress in improving the representation of women of color in technology and computing majors and faculty positions. Although many higher education institutions state that they have made progress in increasing diversity, evidence suggests that there are caveats to this assertion.

Assumption: Increasing Diversity Numbers Equals Success

Much of this report indicates how progress in the advancement, recruitment, and retention has been slow and, in some cases, nonexistent for women of color in tech. Nevertheless, there remains a narrative in higher education mythologizing how progress has indeed occurred, at least for certain groups, with the underlying assumption that an increase in diversity on campus must mean that institutions are taking the right steps. The positivist approach to research has a long history of using numbers to reveal what the researcher deems as the "truth."

The Distinction Between Diversity and Inclusion

Although there have been considerable critiques against positivism (Tuck and McKenzie, 2015), the argument that numbers are best suited to document the totality of experiences for women of color prevails. The presence of this approach promulgates a troubling assumption that precludes thoroughly identifying the complex, interrelated issues affecting women of color in tech.

---

[6] Scholarly devaluation is a component of epistemic exclusion—a theory that "proposes that women and faculty of color are disproportionately harmed by invisible biases built into ostensibly objective and neutral performance standards within systems of evaluation" (Settles et al., 2020, p. 4).

Measuring diversity alone often fails to capture the full picture. Far too often, institutions of higher education measure progress in terms of diversity numbers. Thus, counting and revealing the number of students of color entering and graduating from technology and computing fields has become the norm, with diversity positioned as the only measurement. Stated differently, postsecondary schools recognize and are applauded when they can illustrate an increase of Black and Brown individuals diversifying their student pools or faculty. The committee found this assumption to be problematic for a few reasons.

First, institutions of higher education tend to use "diversity" as the gold star rather than focus on inclusion as a core principle for equity. To assume that a more diverse set of bodies will automatically lead to more equitable ends diminishes the significance of structural barriers and systemic oppression. There are a host of stories from women of color who recognize their staunch oppressors as people belonging to their same race and/or gender social groups. Conversely, there are just as many stories in which women of color attribute their success to mentors who are not women of color. Notably, the committee agreed that there should be more women of color in technology and computing, particularly in leadership roles; however, simply diversifying leadership will not nurture an inclusive environment. Race and gender are not proxies for behavior. Context also involves cultural standards and culture refers to behaviors and expectations. Higher education leaders (e.g., president, provosts, and deans) set the tone for inclusion through their own behaviors and expectations. Without fully unpacking and addressing the complexity of diversity and inclusion, systemic change will remain elusive. The invisibility of microaggressions is what allows white supremacist actions to prevail. Inequity and inequality will be overshadowed by the simple fact that few are willing to name it.

Second, increasing the number of women of color does not necessarily mean a diminishing of the hostile environments they may endure. Without careful attention to understanding the context as well as the variables that inhibit or foster success for women of color, missed opportunities for scaling best practices continue. Take, for example, an analysis by Daily and colleagues (2020), which identified four-year institutions of higher education by various categories (e.g., public, private, non-profit, minority-serving) to chronicle the graduation rate between of various groups of women of color. Using data from IPEDS,[7] their analysis reveals how certain all-women colleges that are predominantly white institutions have impressive rates graduating certain groups of women of color with a computer science degree (Daily et al., 2020).

---

[7] IPEDS is the Integrated Postsecondary Education Data System. It is a system of interrelated surveys conducted annually by the U.S. Department of Education's National Center for Education Statistics (NCES). IPEDS gathers information from every college, university, and technical and vocational institution that participates in the federal student financial aid programs.

Lack of Disaggregated Data

Throughout this report, the committee has discussed the lack of data that are specific to women of color in tech, and more specifically to women of color in tech in the United States. In many cases, data are condensed across categories and presented with all women grouped together or with all women of color grouped together. The small number of women of color is often cited as not sufficient for statistical analysis and serves as the rationale for this aggregation. Privacy concerns are also often cited; however, when these data are not available, women of color lose an opportunity to learn from the experiences of other women who may be facing similar challenges and their strategies for persisting (Begay, 2020; Pawley, 2020). The fear of exposing the few women of color in a technology department or work environment shapes how social science researchers plan and manage their work. When numbers are small enough that women may be identifiable, the push is to aggregate. Stated differently, researchers interested in examining any and all aspects of women of color in higher education typically receive caveats to avoid disaggregating data. The statement "the sample size is too small and risks a reader identifying the subjects" is too often chorused by review committees. "Personally identifiable data" is a phrase meant to protect the few women of color navigating what this report has revealed as hostile non-supportive contexts. But in reality, not reporting the number allows for the maintenance of structures to ignore the issues disproportionately affecting women of color in tech.

Institutions of higher education may also sometimes choose not to disaggregate data based on students' nationalities. As a consequence, non-U.S.-born students of color are often placed within the "underrepresented" category. This strategy leads to claims of "more students of color" entering technology and computing majors, and the fact that the increase is often due to foreign-born students is overlooked. This grouping also obscures where improvements and systemic changes may be needed to attract and retain women of color since non-U.S.-born students of color may need different supports than U.S.-born students or may leave the U.S. tech and computing workforce after completing their education. This narrative presents a partial truth that further troubles the use of the phrase "underrepresented."

IMPROVING THE REPRESENTATION OF
WOMEN OF COLOR IN HIGHER EDUCATION

Although women of color face significant obstacles along the multiple paths they may follow to a career in tech, recognition of these structural barriers provides an opportunity to mitigate or eliminate roadblocks that can stymie their trajectories and implement policies and practices that allow women of color to thrive and reach their full potential. While there is no one-size-fits-all approach to the recruitment, retention, and advancement of women of color in higher education that can be used for all women belonging to any one group, research and practice can elevate promising strategies. An important approach to alter the course for women of color in tech is the creation of inclusive environments supported by policies and practices that are informed by systematic data collection and analysis along with leadership, assessment, and the implementation of effective strategies to promote their advancement (see Chapter 5 for more discussion on federal programs that can shed light on practices to address the underrepresentation of women of color in the sciences).

Provide Early Exposure and Encouragement

In an analysis of experiences of Black women in computing, Ashford (2016) discusses six main themes that emerged in her examination of the experiences of Black women faculty who have persisted in the field of computing education. The first two relate to early exposure, encouragement, and validation:

- "I am set apart": Participants described feeling encouraged and receiving positive reinforcement from parents, teachers, and school administrators. The women also described having been identified as smart and talented in school and set apart in classroom settings as a result. This early reinforcement fostered confidence in their abilities to succeed in computing.
- "I am holding my own": Participants described experiences during their earlier education experiences that allowed them to demonstrate their abilities and where their sense of belonging was positively reinforced.

Along the trajectory from early exposure through career persistence, Ashford's analysis suggests that this exposure and encouragement provides a foundation for women to understand the assets they bring to their work, to make an impact, to feel that they belong and can persist in a doctoral program, and ultimately that they have had significant achievement in their career (Ashford, 2016; Branch, 2020; Gates, 2020; McMullen, 2020).

Lack of family encouragement was mentioned above in this chapter's discussion of assumptions related to why girls and women may not pursue tech degrees. In the face of challenging contexts, parents are an "untapped resource" for recruiting more girls of color into STEM careers in general (Harackiewicz et al., 2012). In her dissertation, Ashley Robinson highlighted the role a mother had in influencing the perception of computing in middle-school African American girls (Robinson, 2015). If a girl's mother had knowledge of information technology careers, she was more likely to consider such a career. Early encouragement notwithstanding, Robinson also reported that a summer workshop exposing middle-school African American girls to human-computer interaction was enough to give participants a positive perception of the discipline and lead them to consider a career in tech (Robinson and Pérez-Quiñones, 2014). The findings in that study are similar to those reported in Denner (2011) where one of the strongest predictors of interest in computing for women was simply "technological curiosity."

Other studies have shown that family support is an influence on Latinx women and women in general (Cohoon, 2001; Denner, 2009). Early findings from Ong and colleagues found that women of color who pursued computing in higher education often entered college with high academic achievement and had early exposure to computing through summer camps, high school programs, or relatives that introduce them to computing. These findings also showed women of color embracing some of the identities that are frequently associated with computing (e.g., that it is for nerds or is geeky), which may help support a sense of belonging (Ong et al., 2020). Therefore, even in the absence of family support, a positive view of the discipline and early encouragement can impact the students' desire to pursue a computing discipline and their ability to persist in tech fields. Higher education institutions have an opportunity to help create opportunities where women of color can engage with computing at an early age—for example, by creating outreach programs, camps during summer or other school vacations, and after-school programs. Such programs can foster positive environments where girls and women of color are able to challenge stereotypical views of success and gain access and experiences in computing prior to entering postsecondary education.

Early encouragement at the undergraduate level in tech disciplines is important in the decision making to continue onto graduate school. The BRAID initiative (Building, Recruiting, And Inclusion for Diversity)[8] is conducting a mixed-methods longitudinal study of computer science departmental changes across 15 institutions. Data are collected from "students, faculty, staff, department chairs, and administrators in order to answer a variety of research questions related to attracting and retaining women and underrepresented minority students in computing majors." Research from this effort has found that building students' academic confidence in introductory courses is key in shaping graduate school plans, but confidence may be moderated by gender, race/ethnicity, or other identities (Wofford, 2021). The BRAID initiative may be a useful evidence-based source for promising institutional practices to increase the number of women of color majoring in computer science.

---

[8] The BRAID initiative is co-led by AnitaB.org and Harvey Mudd College. The effort was launched in September 2014 in partnership with 15 universities across the nation. Since 2014, 15 CS departments ("BRAID Schools") under the leadership of their department chairs have committed to implementing a combination of four commitments in efforts to increase the participation of students from underrepresented groups—racial/ethnic minorities and women—in their undergraduate CS programs. https://anitab.org/braid/ (accessed October 8, 2021).

Pay Attention to Transition Points

As previously discussed, many students face obstacles in their academic career and in particular at transition points—including transitions from community college to a four-year college, from undergraduate to graduate or graduate to faculty, and in promotion and tenure—that can hinder their progress. Women of color also often do not follow the same trajectory to a career in technology that many dominant group members (e.g., white men) follow, for myriad reasons. It is up to academic programs to find a way to increase the support to address the obstacles faced by this group if they want to increase retention with special attention given to the transition points.

Raquel Hill, associate professor and chair of the computer and information sciences department at Spelman College, gave a presentation to the committee in April 2020 highlighting some examples of programming that Spelman has implemented to help equip incoming students with the support they need to thrive and navigate their undergraduate career. For example, Spelman's Women in STEM program is a seven-week summer bridge program for first-year students planning to major in science, engineering, or mathematics. The program includes for-credit coursework, interdisciplinary research projects, enrichment activities, targeted mentoring from student peers and STEM professionals, and academic advising from program staff. The program also covers the full cost of student participation and provides a stipend. Programs such as these help students get hands-on experience in research, engage faculty, and help build relationships between students and faculty (Hill, 2020).

As also discussed above, it is critical to understand the contexts of students and the choices that have shaped their trajectories—for example, whether family obligations, financial issues, or geographic access to community colleges or minority-serving institutions have influenced the track of their academic career (Gates, 2020). It is equally important to recognize that many community colleges and minority-serving institutions are uniquely positioned to provide supports to women of color that can allow them to persist in tech. Tribal colleges and universities, for example, are not only accessible to their communities, but also able to provide students with culturally relevant opportunities and interactions that are targeted to their needs and support their studies. These institutions are also building pathways and partnerships with larger institutions to provide students with opportunities while allowing them to stay in their communities (Baker, 2020).

Transitions from community colleges to four-year institutions or from minority-serving institutions to predominantly white institutions are two further examples of opportunities to target supports for women of color. For example, at Florida International University, the fourth-largest university in the nation and the largest Hispanic-serving institution, 75 percent of students come from community college. Monique Ross, assistant professor in the School of Computing and Information, presented to the committee in April 2020 and provided insights on supports that may help students thrive, such as strengthening peer networks, addressing the digital divide, supporting student organizations that help build community, advocating for diverse hires, providing computing opportunities that are interdisciplinary and allowing students to make a social impact, and funding targeted, culturally relevant programming along the entire academic trajectory from K-12 through career (Ross, 2020). Similar findings were also highlighted in a presentation from Nizhoni Chow-Garcia (2020).

Lastly, academic advisors can play a significant role in the success of students who are women of color, including increasing their sense of belonging (Museus and Ravello, 2010). There are opportunities for academic advisors to facilitate transitions for students at transition points to help women of color successfully navigate higher education systems—especially if they are attuned to the students' needs. More studies are needed to study academic advising using an intersectional lens.

Create Supportive Inclusive Environments

While the number of women of color in academic programs in technology and computing is low, this is not due to lack of qualification or interest in pursuing careers in tech. Reducing obstacles in the academic environment fosters an environment where women of color succeed.

Cultivating Intentional Leadership

This report outlines numerous factors that present challenges for increasing diversity, equity, and inclusion in higher education; however, there are many opportunities to support and drive changes to institutional frameworks and cultures. Institutional leaders have an opportunity to shape the climate of the institutions they lead and to model the inclusivity they are trying to cultivate. During her presentation to the committee, Anna Branch, senior vice president for Equity at Rutgers University, provided questions that the leadership of higher education institutions can consider when evaluating organization culture and climate in order to better understand where exclusionary norms may be impacting women of color. These questions included:

- Who feels included or excluded?
- Who is thriving or not thriving?
- Who is not equitably represented within the institution?
- How do people treat each other?
- What is the historical context for this environment?

Understanding the answers to these questions can help leaders develop intentional strategies (e.g., crediting diversity work toward tenure and promotion or documenting committee service in annual reviews) to shift organizational climate to more inclusive models that can increase the persistence of women of color (Branch, 2020).

It is also important to note that women of color are underrepresented in positions of leadership in higher education institutions. Programs such as HERS[9] and ELATES at Drexel®,[10] which focus on advancing women in higher education leadership, can serve as useful examples in tackling this underrepresentation. Research shows that increased diversity in organizational leadership can increase the sense of belonging and lead to more positive outcomes for women of color, and that organizations and institutions with women of color in the leadership may be more likely to support initiatives that are effective at increasing success for other women of color. However, as discussed earlier in the chapter, women of color of faculty often wear many institutional hats; these initiatives (e.g., networking and creating support networks and cohorts) should come with a caution to not overburden the few women of color who reach leadership roles leading these efforts.

Going Beyond Unconscious Bias Training

Often efforts to encourage the decision makers in higher education to be more culturally responsive and nurture an inclusive environment for women of color to succeed focus on unconscious bias training. While some initiatives can yield positive effects (Moss-Racusin et al., 2016; NRC, 2013), and improvements in awareness of diversity issues and reduced gender bias are heartening outcomes, they are, by themselves, insufficient. Bias trainings will "put folks on notice" of the significance of this topic and educate faculty of risk of misbehavior and benefits. However, effective diversity interventions must also encourage participation by students, faculty, and staff and increase participants' readiness to engage in behaviors that promote gender parity and change the policies and procedures that establish the power structures that sustain these biases. An individual administrator may attend an ally camp, learn to be an effective mentor, identify macroaggressions, and/or explore the distinction between sponsorship and mentoring, yet these approaches fail to consider how power operates *in situ*.

Fostering Supportive Networks and Relationships

Formal and informal networking provides women of color with opportunities to build relationships, share information and experiences, learn about work and research opportunities, and receive guidance from peers and mentors. For example, a lack of senior female mentors can mean that junior faculty have less guidance on the unstated rules for promotion and tenure hindering their chances

---

[9] For more information see https://www.hersnetwork.org/programs/overview/ (accessed October 7, 2021).

[10] For more information see https://drexel.edu/provost/initiatives/elates/about/ (accessed October 7, 2021).

of moving upward (NRC, 2013). In addition to helping students and faculty navigate the higher education environment, mentors and peers can help women of color recognize the assets they bring to the table—cultural capital, for example. These networks (such as student groups for women of color and campus chapters of professional organizations) and relationships (such as tenured faculty role models) allow women of color to build social capital and feel confidence in their competencies that can help them thrive and persist in tech and provide safe spaces where they feel supported (Carpenter, 2020; McMullen, 2020; Ong et al., 2020; NRC, 2013). Furthermore, the gender of a role model appears to be less important than the person's ability to challenge stereotypes. The counter-stereotypical male role model can be just as helpful as a female role model in promoting women's beliefs about success in STEM (Cheryan et al., 2011). Encouraging the adoption of counter-stereotypical signals among male faculty may be an actionable step to help foster women's self-confidence in STEM (Charlesworth and Banaji, 2019). Ultimately, students and faculty from underrepresented groups who do not have access to these types of mentorship and/or guidance are more likely to feel excluded by peers in tech and other STEM environments, and ultimately leave the sciences (NASEM, 2019a; NRC, 2013).[11]

Beena Sukumaran, professor of civil and environmental engineering and vice president for research at Rowan University, provided the committee with numerous examples at its May 2020 workshop of the holistic approach that the College of Engineering at Rowan University is taking to improve diversity, equity, and inclusion (DEI) practices to help students thrive. Practices implemented as part of this approach included changing admissions standards (e.g., changing the evaluation process for transfer students, making the SAT optional for admission); increasing understanding of DEI among students, faculty, and administrators to create a culture of "collective intentionality" across departments; developing an "Advocates and Allies" program for incoming freshmen and transfer students to facilitate transition, retention, and graduation; transforming existing sophomore and junior level curriculum and providing faculty with guidance on how to design an inclusive curriculum; and working to strengthen students' aspirations and identities as engineers by inviting speakers who are "role models of difference" and who have impacted society and policy in their professional careers (Sukumaran, 2020).

In 2013, testimony from academic women of color in computing, heard at a National Academies of Sciences, Engineering, and Medicine meeting, urged institutions to provide structures for mentoring and "provide resources for establishing virtual and in-person networks of academic women of color in computing to allow for the needed "sticking together" and "blending in"[12] mentoring, the

---

[11] Mentoring underrepresented students in STEMM: A Survey and Discussion: https://www.nap.edu/resource/25568/McGee%20-%20STEMM%20Mentoring%20Identity.pdf.

[12] "Blending in" means the woman of color would develop a network that includes those with power, which are often white and/or male, while "sticking together" means the woman of color would create a network consisting of those similar to herself (NRC, 2013).

sharing of best practices, and for senior academic women of color in computing to be visible role models to junior academic women of color as well as women students of color" (NRC, 2013). The push for virtual communities in fostering networks for women of color is especially salient now given the remote environment created by COVID-19 where students and faculty are less physically tied to their campus environments. One example is the non-profit organization *Rewriting the Code*,[13] a virtual community whose mission is to support college, graduate, and early career women in tech through intersectional communities, mentorship, industry experience, and educational resources. The organization has created subgroups that members may join, including for Black women, Latinx women, and international students, that have a stated goal to build a sense of inclusion and belonging by offering a place for sharing experiences and advice with a familiar community.

It is also worth reiterating that best practices can be developed from further study into why certain sub-disciplines in computing have been better at mentoring and supporting women, and whether strategies that attract women of color to these sub-disciplines could be applied to other technical sub-disciplines where women of color are less represented.

Cultivating Partnerships to Develop and Implement Best Practices

A number of speakers at the committee's public workshops discussed the value of learning from peer institutions, colleagues, and communities that have created environments, programs, and practices that support the success of women of color in tech disciplines. Many minority-serving institutions have created strong computing programs that are graduating women of color at significantly higher rates than predominantly white institutions. Leadership at colleges and universities that are looking to improve outcomes can learn from other institutions that are succeeding to learn more about the policies, programs, and climate that foster success for women of color (Adams, 2020). Ashley Carpenter, assistant professor at Appalachian State University,[14] also underscored the role that minority-serving institutions can play by sharing insights on how to support women of color. Carpenter encouraged higher education institutions to work to build supportive and safe communities of practice on campus, provide career guidance and professional development, and facilitate connections with potential mentors (Carpenter, 2020; Pinkston, 2020). HBCUs, for example, have shown disproportionate success in graduating African American students, particularly in the STEM fields, which has been attributed in part to their strong academic and social support networks and culturally responsive teaching approaches (NASEM,

---

[13] For more information see https://rewritingthecode.org/ (accessed October 7, 2021).

[14] Carpenter is listed in the workshop meeting agenda found in Appendix C with her previous affiliation as University Center for Exemplary Mentoring and Diversity Initiatives Program Coordinator at Massachusetts Institute of Technology.

2019a, 2019b). However, it is important to note that HBCUs are not devoid of structural barriers to the success of women of color in technology and computing. Given the low numbers of female faculty and women of color faculty in computing, campus and departmental climate issues can still persist.

Joan Reede, dean for diversity and community partnership at Harvard University Medical School, presented to the committee in February 2020 and underscored the importance of leadership in shaping organizational culture. Reede highlighted the importance of integrating diversity and inclusion into the mission and values of institutions and leadership that works to ensure that efforts to increase success of women of color are adequately resourced, consistent, and long enough for sustainable outcomes to be achieved. Reede pointed to examples of how Harvard Medical School is working within communities as well as within the institution to create opportunities for future student success—for example, by offering out-of-school programming to high school students during the academic year and summer and offering supplemental programming to high schools that do not have or cannot offer advanced placement coursework. For students already at Harvard, opportunities exist to participate in summer research programs, externships, and postdoctoral fellowships. Commitment as an institution to prioritizing continuity of programming to foster development along the academic trajectory is helping to create multiple points of access to opportunities for women of color.

Regarding partnerships between higher education and industry, presenters to the committee discussed the importance of symbiotic relationships. Higher education institutions have a unique opportunity in these partnerships not only to learn from industry and help to shape curriculum that can help students to succeed, but also to help equip industry with a better understanding of their students and the training and supports needed to help women of color successfully transition from higher education into the workforce (Gates, 2020; Washington, 2020).

Accountability

Systems of accountability play a critical role in setting institutional goals for attracting and retaining women of color as students and faculty. Leadership at all institutional levels needs metrics in order to create strategic goals to strive for and to track, monitor, and evaluate whether those efforts have been successful (Adams, 2020; Reede, 2020). Adequate data collection and disaggregation plays a critical role in increasing transparency and informing evidence-based decisions. Kaye Husbands Fealing, chair and professor in the School of Public Policy at Georgia Tech, presented data from Leggon (2018), which highlights evaluation criteria for effective initiatives that increase participation and improve target group experiences and which can be used at the program and institutional level, initiatives such as increasing the diversity of the professoriate, developing culturally inclusive curricula, or increasing community engagement (Husbands Fealing, 2020) (Table 3-1).

TABLE 3-1 Evaluation Criteria for Effective Initiatives

Program Level	Institutional Level
• Clearly defined goals, objectives, priorities, and outcomes	• Institutionalization: sustained commitment, bottom-up and top-down
• Education, training, and socialization	• Integrated organizational strategy embedded into the basic structure, strategy, and standard operating procedures of the organization
• Networking and community building—creating a sense of belonging	
• Mentor-protégé programs	
• Formative evaluation and continuous improvement	• Management accountability and evaluation
• Longitudinal tracking	
• Bridge mechanisms—one program to another, one level to the next (e.g., from the program level to the institutional level), across sectors	

SOURCE: Leggon (2018).

Widen Recruitment Efforts

Changing recruitment efforts to access a broader pool of candidates is a key strategy for addressing the low numerical representation of women in STEM fields. Increasing the number of women in these fields can attract future generations of girls and women to STEM. Broadening the pool of STEM talent would help address the shortage of qualified STEM educators, which, in turn, would introduce more students to and increase students' excitement about STEM topics (Diekman et al., 2017).

Evidence suggests that in many STEM fields, disparities in the number of women in STEM tenure-track positions cannot be explained by a lack of qualified candidates. Even after earning a doctoral degree in a STEM field, more women than men opt out of applying for research-intensive academic positions. One study found that, across six scientific disciplines, the proportion of female degree holders was larger than the proportion of female applicants for research-intensive tenure-track positions (NRC, 2010). An analysis of approximately 3,000 faculty from 14 universities (Kaminski and Geisler, 2012) found that although men are more likely to be hired into faculty positions, women who are hired tend to be retained at similar rates to men. Therefore the authors of this analysis concluded that when women are hired they are likely to persist. Thus, the key stage to increase representation of women in academic positions in the sciences appears to be during recruitment (Diekman et al., 2017).

Juan Gilbert, professor and chair of the Computer and Information Science and Engineering Department at the University of Florida, provided insights to the committee in July 2020 on effective strategies for proactively increasing the recruitment of women of color at both the student and faculty level (Gilbert, 2020). Gilbert's approach to proactive recruiting, which is grounded in research and practice, emphasizes a number of methods for improving outcomes for women of color by leveraging innovative recruitment methods and implementing strategies to build support from leadership, foster community, provide guidance and mentorship, and promote professional development. In examining the characteristics of groups who were persisting both as students and faculty, Gilbert noted strategies for success and applied them to his own efforts to increase diversity in computing. A primary strategy is building communities of practice by recruiting in cohorts. In his own work, Gilbert had seen how peers with similar experiences were able to provide support and strategic guidance that helped members of their cohort to persist in the field. This was true both at the student and the faculty level. In cases where a single department does not have the resources to hire a cohort, higher education institutions can facilitate the coordination of hiring across departments or colleges to create cohorts of women of color who can support one another across the institution.

When cohort hiring is not possible at all, institutions can still leverage faculty of color to create peer support networks. Gilbert also highlighted the possibility of expanding the types and number of technology and computing disciplines to increase diversity. As previously discussed in this chapter, there may be more women of color in specific computing sub-disciplines. Institutions with the capacity to do so could consider adding new areas of research to their computing departments in order to expand opportunities to recruit more women of color into technology and computing fields (Gilbert, 2020).

Naturally, in the case of student admissions and hiring of faculty, the question arises of how to recruit cohorts of qualified women of color who, like all applicants, must compete for a limited number of openings. Although many institutions have undertaken efforts to diversify slates of candidates, personal networks and recommendations from non-diverse hiring committees or admissions panels often influence the demographic composition of the groups of candidates that are considered. Gilbert discussed one strategy to address this. "Application's Quest," a technology he has invented and patented uses artificial intelligence to holistically evaluate and compare applications with the goal of diversifying recommendations for admissions and hiring by reducing bias. This type of holistic evaluation broadens networks for recruitment and increases the diversity of qualified candidate pools from which to admit students or hire faculty. Gilbert has found that the results of using this technology is a slate of diverse candidates with equal achievement outcomes without a reduction in candidate achievement outcomes (Gilbert, 2020). Furthermore, research on the rates at which women of color faculty are offered tenure vs. non-tenure track positions, startup packages, and competitive salaries, among others recruitment factors, would help to build evidence on how best to recruit this group.

It is important to note that while surging enrollments in computer science can be seen as positive growth, they can also trigger institutions to put screens in place to manage enrollment demands which has the potential to negatively impact diversity in tech disciplines. This reasoning has been used to explain the sudden drop in the participation of women in computing in the 1980s. It has been proposed that the practices put in place at many universities created an inhospitable and sometimes hostile climate that female students found to be uninviting and off-putting (CRA, 2015; NASEM, 2018b).

Leadership plays an important role in providing the financial and institutional support needed to implement these policies, demonstrating a willingness to shift the paradigm of how institutions of higher education adopt practices to increase the recruitment and retention of women of color, both as students and as faculty.

Capture the Experiences of Women of Color in Higher Education

An overarching finding in the committee's examination of the research literature and evidence presented by experts at the committee's series of public workshops is the need for data related to women of color that is disaggregated on a number of dimensions in order to better understand both their unique and shared experiences. Although the committee understands the challenges and privacy concerns that must be considered when using small datasets, there are a multitude of examples of other qualitative research methodologies that are valid alternatives to collect data on small sample sizes, such as case studies and autoethnographies. Course evaluations are another example: Institutions allow class evaluations no matter how small the sample size. Seminars with a handful of people are encouraged to be assessed. Granted, these data cannot be published, but they are often used to inform decisions on important issues such as promotion and tenure.

The same reasoning is not applied to data collection to understand the experiences of women of color in tech. However, defining and describing the problem need not be tethered to statistical significance and quantitative research—discovering the problem is more significant than replicability. In software testing one can prove the absence of bugs, but not correctness. The statement that the population of women of color in technology and computing fields is too small to study is an excuse to define a single case; however, examining qualitative data that provide more insights into the experiences of women of color can inform strategic efforts to change trajectories for student and faculty women of color using evidence-based practices, despite the limited sample size. For example, a recent paper from the Center for Inclusive Computing (CIC)[15] used information from 22 site visits to identify the data to collect and questions to ask to help a university self-diagnose why equitable representation remains elusive. These questions may be helpful in demonstrating how higher education institutions can go beyond reporting on just diversity numbers to better understand structural

---

[15] CIC is based at Northeastern University awards grants to colleges and universities to support the implementation of evidence-based approaches that increase the representation of women in computing. For more information see https://cic.khoury.northeastern.edu (accessed October 7, 2021).

barriers and systemic oppression and enable leaders to implement well-informed broadening participation strategies (Brodley et al., 2021).

RECOMMENDATIONS

The committee offers the following recommendations regarding the recruitment, retention, and advancement of women of color in higher education.

RECOMMENDATION 3-1. To foster continuous pathways for women of color in higher education, institutions at the departmental, college, and university levels should promote the collection of empirical qualitative and quantitative data that disaggregate the recruitment and graduation experiences of students, the recruitment and promotion and tenure trajectories of all faculty, and ascension to leadership positions for women of color.

These data should be used to inform the design and implementation of the following processes, but not limited to

- Culturally responsive review processes of promotion and tenure guidelines and academic review processes to ensure that the qualitative and quantitative research produced by women of color in tech is equally valued at the departmental, college, and university levels.
- Collection, analysis, and presentation of disaggregated data of tech departments and college environments to institutional leaders. Information regarding the individuals who constitute research teams, laboratories, faculty service committees, and doctoral committees could be used to determine whether one group is disproportionately receiving opportunities or assuming invisible labor. These data should also include the social categories (e.g., race/ethnicity, gender, socioeconomic status) of decision makers at departmental, college, and university levels in order to understand how power operates as an intersectional concept.
- A reward system sustained by computing and other technology-related departments and college environments which demonstrate ongoing levels of success recruiting, retaining, and maintaining an inclusive context for women of color in tech. Both disaggregated qualitative and quantitative could be used to present cases that illustrate effective strategies.
- Programs, policies, and practices that may be tailored to support the recruitment, retention, and advancement of different groups of women of color.

RECOMMENDATION 3-2. Institutions of higher education should collect and analyze disaggregated qualitative data to document the voices of women of color in tech and the narrated experiences of those who work with women of color that demonstrate how women of color fare in technology and computing courses as they navigate higher education at various levels.

To accomplish this, leaders in higher education, such as provosts, deans, and department heads, should use these data as the basis for their decisions for developing, sourcing, and evaluating initiatives for students and faculty who are women of color. Leaders in higher education should

- Regularly review and interpret these narrative data as barometers for measuring progress toward diversity, equity, and inclusion goals, and
- Identify and adopt best practices from institutions that have successfully recruited and retained women of color in tech.

RECOMMENDATION 3-3. Higher education leaders should widen recruitment efforts to identify women of color candidates to join their computer science, computer engineering, and other tech departments as students and faculty, with increased consideration of those from two-year community colleges and minority-serving institutions, and should develop retention strategies focused on supporting these students and faculty during transitions to their institutions.

Strategies should include the following:

- Developing partnerships with two-year community colleges and minority-serving institutions to identify and recruit tech students and graduates who are women of color.
- Increase access to higher education by integrating financial assistance programs with recruitment and retention strategies that target undergraduate and graduate students who are women of color.
- Providing increased social supports for incoming tech students and faculty who are women of color, such as orientations, professional development, career coaching, and peer mentoring. Individuals who provide this support should be required to maintain ongoing, regular training in culturally responsive education, racial awareness, and intersectionality.

REFERENCES

ACE (American Council for Education). 2021. International Briefs for Higher Education Leaders: Women's Representation in Higher Education Leadership Around the World. Vol 9. Washington, DC: American Council for Education.

Adams, S. 2020. Underrepresentation of Women of Color in Tech: Higher Education and Academia." Presentation made at the public workshop of the Committee on Addressing the Underrepresentation of Women of Color in Tech, Washington, DC, February 2020.

Adcock, T. 2014. Technology integration in American Indian education: An overview. *Journal of American Indian Education* 104-121.

Ashcraft, C., B. McLain, and E. Eger. 2016. *Women in tech: The facts.* Boulder, CO: National Center for Women and Information Technology. https://ncwit.org/resource/thefacts/.

Ashford, S. N. 2016. Our counter-life herstories: The experiences of African American women faculty in US computing education. Doctoral dissertation. University of South Florida.

Baker, T. 2020. Women of Color in Tech: A Focus on Native Americans. Presentation made at the public workshop of the Committee on Addressing the Underrepresentation of Women of Color in Tech, virtual, June 2020. https://www.nationalacademies.org/event/06-04-2020/addressing-the-underrepresentation-of-women-of-color-in-tech-workshop-4-day-2.

Begay, S. 2020. The Current State of STEM Education for American Indian and Alaska Native Communities. Presentation made at the public workshop of the Committee on Addressing the Underrepresentation of Women of Color in Tech, virtual, June 2020. https://www.nationalacademies.org/event/06-04-2020/addressing-the-underrepresentation-of-women-of-color-in-tech-workshop-4-day-2.

Bobb, K. 2020. Panel 1 Session. Presentation made at the public workshop of the Committee on Addressing the Underrepresentation of Women of Color in Tech, virtual, May 2020. https://www.nationalacademies.org/event/05-14-2020/addressing-the-underrepresentation-of-women-of-color-in-tech-workshop-day-1.

Branch, A. 2020. When and Where I Enter? Exits, Pathways, and Potholes for Black Women in Tech. Presentation made at the public workshop of the Committee on Addressing the Underrepresentation of Women of Color in Tech, virtual, April 2020. https://www.nationalacademies.org/event/04-08-2020/addressing-the-underrepresentation-of-women-of-color-in-tech-day-2.

Brodley, C. E., C. Gill, and S. Wynn. 2021. Diagnosing why Representation Remains Elusive at your University: Lessons Learned from the Center for Inclusive Computing's Site Visits. Respect 2021 Conference. http://respect2021.stcbp.org/wp-content/uploads/2021/05/102_Experience-Report_02_paper_13.pdf.

Canning, E. A., K. Muenks, D. J. Green, and M. C. Murphy. 2019. STEM faculty who believe ability is fixed have larger racial achievement gaps and inspire less student motivation in their classes. *Science Advances* 5(2):eaau4734. https://advances.sciencemag.org/content/5/2/eaau4734.

Carpenter, A. 2020. Addressing the Underrepresentation of Women of Color in Technology. Presentation made at the public workshop of the Committee on Addressing the Underrepresentation of Women of Color in Tech, virtual, May 2020. https://www.nationalacademies.org/event/05-14-2020/addressing-the-underrepresentation-of-women-of-color-in-tech-workshop-day-1.

Cech, E. A., and M. Blair-Loy. 2010. Perceiving glass ceilings? Meritocratic versus structural explanations of gender inequality among women in science and technology. *Social Problems* 57(3):371-397.

Charlesworth, T. E. S., and M. R. Banaji. 2019. Gender in science, technology, engineering, and mathematics: Issues, causes, solutions. *Journal of Neuroscience* 39(37):7228-7243.

Chow-Garcia, 2020. N. Native American Women and Indigenous Models for Success. Presentation made at the public workshop of the Committee on Addressing the Underrepresentation of Women of Color in Tech, virtual, June 2020. https://www.nationalacademies.org/event/06-04-2020/addressing-the-underrepresentation-of-women-of-color-in-tech-workshop-4-day-2.

Clauset, A., S. Arbesman, and D. B. Larremore. 2015. Systematic inequality and hierarchy in faculty hiring networks. *Science Advances* 1(1):e1400005.

Cohoon, J. M. 2001. Toward improving female retention in the computer science major. *Communications of the ACM* 44(5):108-114.

Cohoon, J. M., S. Nigai, and J. Kaye. 2011. Gender and computing conference papers. *Communications of the ACM* 54(8):72-80. https://doi.org/10.1145/1978542.1978561.

Collins, P. H. 2019. *Intersectionality as critical social theory.* Durham, NC: Duke University Press.

Corbett, C., and C. Hill. 2015. *Solving the equation: The variables for women's success in engineering and computing.* Washington, DC: American Association of University Women. https://www.aauw.org/app/uploads/2020/03/Solving-the-Equation-report-nsa.pdf.

CRA (Computing Research Association). 2015. Expanding the Pipeline: Booming Enrollments – What is the Impact? *Computing Research News* 27(5). https://cra.org/crn/wp-content/uploads/sites/7/2015/07/CRN_May_2015.pdf.

Crenshaw, K. 1991. Mapping the margins: Identity politics, intersectionality, and violence against women. *Stanford Law Review* 43(6):1241-1299.

Daily, S., W. Eugene, C. Shelton, and J. Thomas. 2020, August 27. Trends in Bachelors among Women of Color in Computing. Women of Color in Computing Conference, Virtual.

DeerInWater, K. 2020. Women of Color in Tech: A Focus on Native Americans in Computing. Presentation made at the public workshop of the Committee on Addressing the Underrepresentation of Women of Color in Tech, virtual, June 2020. https://www.nationalacademies.org/event/06-04-2020/addressing-the-underrepresentation-of-women-of-color-in-tech-workshop-4-day-2.

Delgado-Olson, A. 2020. Addressing Underrepresentation of Women of Color in Tech: Perspectives of Native American Women in Computing. Presentation made at the public workshop of the Committee on Addressing the Underrepresentation of Women of Color in Tech, virtual, June 2020. https://www.nationalacademies.org/event/06-04-2020/addressing-the-underrepresentation-of-women-of-color-in-tech-workshop-4-day-2.

Denner, J. 2009. The role of the family in the IT career goals of middle school Latinas. AMCIS 2009 Proceedings, p. 334.

Denner, J. 2011. What predicts middle school girls' interest in computing? *International Journal of Gender, Science and Technology* 3(1):54-69.

Diekman, A. B., E. S. Weisgram, and A. L. Belanger. 2015. New routes to recruiting and retaining women in STEM: Policy implications of a communal goal congruity perspective. *Social Issues and Policy Review* 9:52-88. http://doi.org/10.1111/sipr.12010.

Diekman, A. B., M. Steinberg, E. R. Brown, A. L. Belanger, and E. K. Clark. 2017. A goal congruity model of role entry, engagement, and exit: Understanding communal goal processes in STEM gender gaps. *Personality and Social Psychology Review* 21(2):142-175.

Dotson, K. 2011. Tracking epistemic violence, tracking practices of silencing. *Hypatia* 26(2):236-257.

Dotson, K. 2012. A cautionary tale: On limiting epistemic oppression. *Frontiers: A Journal of Women Studies* 33(1):24-47.

Dotson, K. 2014. Conceptualizing epistemic oppression. *Social Epistemology* 28(2):115-138.

Funk, C., and K. Parker. 2018. *Women and Men in STEM Often at Odds over Workplace Equity.* Pew Research Center.

Gates, A. 2020. Underrepresentation of Women of Color in Tech: Higher Education and Academia. Presentation made at the public workshop of the Committee on Addressing the Underrepresentation of Women of Color in Tech, Washington, DC, February 2020.

Gilbert, J. E. Addressing the Underrepresentation of Women of Color in Tech. Presentation made to the Committee on Addressing the Underrepresentation of Women of Color in Tech, virtual, July 2020.

Glass, J. L., S. Sassler, Y. Levitte, and K. M. Michelmore. 2013. What's so special about STEM? A comparison of women's retention in STEM and professional occupations. *Social Forces* 92(2):723-756.

Harackiewicz, J. M., C. S. Rozek, C. S. Hulleman, and J. S. Hyde. 2012. Helping parents to motivate adolescents in mathematics and science: An experimental test of a utility-value intervention. *Psychological Science* 23(8):899-906.

Hill, C., K. Miller, K. Benson, and G. Handley. 2016. Barriers and Bias: *The Status of Women in Leadership*. American Association of University Women.

Hill, R. 2020. Retention of Women of Color Students in Tech: An MSI Perspective. Presentation made at the public workshop of the Committee on Addressing the Underrepresentation of Women of Color in Tech, virtual, April 2020. https://www.nationalacademies.org/event/04-07-2020/addressing-the-underrepresentation-of-women-of-color-in-tech-day-1.

Holloway, B. M., T. Reed, P. K. Imbrie, and K. Reid. 2014. Research-informed policy change: A retrospective on engineering admissions. *Journal of Engineering Education* 103(2):274-301.

Hunt, D. I., and B. Nicodemus. 2014. Gatekeeping in ASL-English interpreter education programs: Assessing the suitability of students for professional practice. In D. J. Hunt and S. Hafer (Eds.) *Proceedings from CIT 2014: Our roots: The essence of our future*. Pp. 44-60. Portland, OR: Conference of Interpreter Trainers.

Hunt, J. 2010. Why do women leave science and engineering? NBER working paper no. 15853. Cambridge, MA: National Bureau of Economic Research.

Hurtado, S., D.F. Carter, and A. Spuler. 1996. Latino student transition to college: Assessing difficulties and factors in successful college adjustment. *Research in Higher Education* 37(2):135-157.

Hurtado, S., O. S. Cerna, J. C. Chang, V. B. Saenz, L. R. Lopez, C. Mosqueda, L. Oseguera, M. J. Chang, and W. S. Korn. 2006. *Aspiring scientists: Characteristics of college freshmen interested in the biomedical and behavioral sciences*. Los Angeles, CA: Higher Education Research Institute. https://heri.ucla.edu/PDFs/NIH/Summer%20Report.PDF.

Hurtado, S., J. C. Han, V. B. Sáenz, L. L. Espinosa, N. L. Cabrera, and O. S. Cerna. 2007. Predicting transition and adjustment to college: Biomedical and behavioral science aspirants' and minority students' first year of college. *Research in Higher Education* 48(7):841-887.

Husbands Fealing, K. 2020. What Can Leadership Do? Effective practices within organizations to retain and advance women of color in STEM. Presentation made at the public workshop of the Committee on Addressing the Underrepresentation of Women of Color in Tech, virtual, April 2020. https://www.nationalacademies.org/event/04-08-2020/addressing-the-underrepresentation-of-women-of-color-in-tech-day-2.

Jaschik, S. 2010. *Different paths to full professor*. Inside Higher Ed. https://www.insidehighered.com/news/2010/03/05/different-paths-full-professor.

Jaschik, S. 2016. *Unintended Help for Male Professors*. Inside Higher Ed. https://www.insidehighered.com/news/2016/06/27/stopping-tenure-clock-may-help-male-professors-more-female-study-finds.

Kaminski, D., and C. Geisler .2012. Survival analysis of faculty retention in science and engineering by gender. *Science* 335:864-866. doi:10.1126/science.1214844.

Kang, S. K., and S. Kaplan. 2019. Working toward gender diversity and inclusion in medicine: Myths and solutions. *The Lancet* 393(10171):579-586.

Kidd, I. J., and H. Carel. 2017. Epistemic injustice and illness. *Journal of Applied Philosophy* 34(2):271-190.

Kim, T., D. Peck, and B. Gee. 2018. *Race, Gender & the Double Glass Ceiling: An Analysis of EEOC National Workforce Data*. The Ascend Foundation.

Leggon, C. B. 2018. Reflections on broadening participation in STEM: What do we know? What do we need to know? Where Do We Go From Here? *American Behavioral Scientist* 62(5):719-726.

Malcom, L. E., and S. M. Malcom. 2011. The double bind: The next generation. *Harvard Educational Review* 81(2):162-172.

Malcom, S. M., P. Q. Hall, and J. W. Brown. 1976. *The double bind: The price of being a minority woman in science*. Publication 76-R-3. Washington, DC: American Association for the Advancement of Science. http://web.mit.edu/cortiz/www/Diversity/1975-DoubleBind.pdf.

McAlear, F., A. Scott, K. Scott, and S. Weiss. 2018. Women of color in computing. Data brief.

McCullough, T. 2016. The myth women in tech need to stop believing. *Forbes*, February 6. https://fortune.com/2016/02/06/myth-women-tech/.

McGee, E. O. 2020. Interrogating structural racism in STEM higher education. *Educational Researcher* 49(9):633-644.

McGee, E. O. 2021. The agony of stereotyping holds Black women back. *Nature Human Behaviour*, 5(1), 3.

McMullen, K. 2020. Increasing the Representation of Women of Color in Technology. Presentation made at the public workshop of the Committee on Addressing the Underrepresentation of Women of Color in Tech, Washington, DC, February 2020.

Metoyer, R., M. A. Pérez-Quiñones, A. Bazerianos, and J. Woodring. 2019. Retention. In *Diversity in visualization*, edited by R. Metoyer and K. Gaiter. San Rafael, CA: Morgan and Claypool Publishers. Pp. 39-52. https://www.morganclaypool.com/doi/10.2200/S00894ED1V01Y201901VIS010.

Moss-Racusin, C. A., J. F. Dovidio, V. L. Brescoll, M. J. Graham, and J. Handelsman. 2012. Science faculty's subtle gender biases favor male students. *Proceedings of the National Academy of Sciences of the United States of America* 109(41):16474-16479.

Moss-Racusin, C. A., J. van der Toorn, J. F. Dovidio, V. L. Brescoll, M. J. Graham, and J. Handelsman. 2016. A "scientific diversity" intervention to reduce gender bias in a sample of life scientists. *CBE—Life Sciences Education* 15(3):ar29.

Museus, S. D., and J. N. Ravello. 2010. Characteristics of academic advising that contribute to racial and ethnic minority student success at predominantly white institutions. *NACADA Journal* 30(1):47-58.

NAS, NAE, and IOM (National Academy of Sciences, National Academy of Engineering, and Institute of Medicine). 2011. *Expanding underrepresented minority participation: America's science and technology talent at the crossroads*. Washington, DC: The National Academies Press. https://doi.org/10.17226/12984.

NASEM (National Academies of Sciences, Engineering, and Medicine). 2013. *Seeking solutions: Maximizing American talent by advancing women of color in academia*. Washington, DC: The National Academies Press.

NASEM. 2016. *Barriers and opportunities for 2-year and 4-year STEM degrees: Systemic change to support students' diverse pathways*. Washington, DC: The National Academies Press. https://doi.org/10.17226/21739.

NASEM. 2018a. *Sexual harassment of women: Climate, culture, and consequences in academic sciences, engineering, and medicine*. Washington, DC: The National Academies Press. https://doi.org/10.17226/24994.

NASEM. 2018b. *Assessing and responding to the growth of computer science undergraduate enrollments*. Washington, DC: The National Academies Press. doi: 10.17226/24926.

NASEM. 2019a. *The science of effective mentorship in STEMM*. Washington, DC: The National Academies Press. https://doi.org/10.17226/25568.

NASEM. 2019b. *Minority serving institutions: America's underutilized resource for strengthening the STEM workforce*. Washington, DC: The National Academies Press. https://doi.org/10.17226/25257.

NASEM. 2020. *Promising practices for addressing the underrepresentation of women in science, engineering, and medicine: Opening doors*. Washington, DC: The National Academies Press. https://doi.org/10.17226/25585.

NCES (National Center for Education Statistics). 2017. Beginning college students who change their majors within 3 years of enrollment. *Data Point*, NCES 2018-43. Washington, DC: Institute of Education Sciences, Department of Education. https://nces.ed.gov/pubs2018/2018434.pdf.

NCES. 2020. Web Tables. *A 2017 follow-up: Six-year persistence and attainment at first institution for 2011-12 first-time postsecondary students*. Washington, DC: Institute of Education Sciences, Department of Education. https://nces.ed.gov/pubs2020/2020237.pdf.

NEA (National Education Council). 2021. *Beyond the Bubble: Americans Want Change on High Stakes Assessments*. National Education Council.

NRC (National Research Council). 2010. *Gender differences in critical transitions in the careers of science, engineering, and mathematics faculty.* Washington, DC: The National Academies Press. doi:10.17226/12062

NRC. 2013. *Seeking solutions: Maximizing American talent by advancing women of color in academia: Summary of a conference.* Appendix A-2: Women of color among STEM faculty: Experiences in academia. Washington, DC: The National Academies Press. https://doi.org/10.17226/18556.

Ong, M., N. Jaumot-Pascual, and L. T. Ko. 2020. Research literature on women of color in undergraduate engineering education: A systematic thematic synthesis. *Journal of Engineering Education* 109(3):581-615.

Pawley, A. 2020. Learning from Small Numbers: Theory Informed Insights on Gender and Race. Presentation made at the public workshop of the Committee on Addressing the Underrepresentation of Women of Color in Tech, virtual, June 2020. https://www.nationalacademies.org/event/06-04-2020/addressing-the-underrepresentation-of-women-of-color-in-tech-workshop-4-day-2.

Peck, D. 2020. An Analysis of the Intersection of Race and Gender in San Francisco Bay Area Technology Workforce 2007-2015. Presentation made at the public workshop of the Committee on Addressing the Underrepresentation of Women of Color in Tech, virtual, May 2020.

Pinkston, T. 2020. The Parity Objective: Developing a 'Framework' Suite of Best Practices Towards Achieving It. Presentation made at the public workshop of the Committee on Addressing the Underrepresentation of Women of Color in Tech, virtual, May 2020. https://www.nationalacademies.org/event/05-14-2020/addressing-the-underrepresentation-of-women-of-color-in-tech-workshop-day-1.

Reede, J. 2020. Moving Beyond a Seat at the Table: Underrepresentation of Women of Color in Higher Education and Academia. Presentation made at the public workshop of the Committee on Addressing the Underrepresentation of Women of Color in Tech, Washington, DC, February 2020.

Reyes, M. E. 2011. Unique challenges for women of color in STEM transferring from community colleges to universities. *Harvard Educational Review* 81(2):241-263.

Riegle-Crumb, C., B. King, and Y. Irizarry. 2019. Does STEM stand out? Examining racial/ethnic gaps in persistence across postsecondary fields. *Educational Researcher* 48(3):133-144.

Robinson, A. 2015. The attitudes of African American middle school girls toward computer science: Influences of home, school, and technology use. PhD Dissertation. Virginia Tech.

Robinson, A., and M. A. Pérez-Quiñones. 2014. Underrepresented middle school girls: On the path to computer science through paper prototyping. *Proceedings of the 45th ACM Technical Symposium on Computer Science Education*, pp. 97-102. https://doi.org/10.1145/2538862.2538951.

Ross, M. 2020. Retention of Women of Color Students in Tech: An MSI Perspective. Presentation made at the public workshop of the Committee on Addressing the Underrepresentation of Women of Color in Tech, virtual, April 2020. https://www.nationalacademies.org/event/04-07-2020/addressing-the-underrepresentation-of-women-of-color-in-tech-day-1.

Rosser, V. J. 2004. Faculty member' intentions to leave: A national study on their worklife and satisfaction. *Research in Higher Education* 45(3):285.

Ryan, M. K., Haslam, S. A., Morgenroth, T., Rink, F., Stoker, J., and K. Peters. 2016. Getting on top of the glass cliff: Reviewing a decade of evidence, explanations, and impact. *The Leadership Quarterly* 27(3):446-455.

Settles, I. H., M. K. Jones, N. T. Buchanan, and K. Dotson. 2020. Epistemic exclusion: Scholar(ly) devaluation that marginalizes faculty of color. *Journal of Diversity in Higher Education.* https://doi.org/10.1037/dhe0000174.

Steele, C. M. 1997. A threat in the air: How stereotypes shape intellectual identity and performance. *American Psychologist* 52(6):613.

Su, X., J. Johnson, and B. Bozeman. (2015). Gender diversity strategy in academic departments: Exploring organizational determinants. *Higher Education* 69(5):839-858.

Sukumaran, B. 2020. Revolutionizing Engineering and Technology Diversity with a focus on Undergraduate Education. Presentation made at the public workshop of the Committee on Addressing the Underrepresentation of Women of Color in Tech, virtual, May 2020. https://www.nationalacademies.org/event/05-14-2020/addressing-the-underrepresentation-of-women-of-color-in-tech-workshop-day-1.

Thompson, M., and D. Sekaquaptewa. 2002. When being different is detrimental: Solo status and the performance of women and racial minorities. *Analyses of Social Issues and Public Policy* 2(1):183-203.

Tuck, E., and M. McKenzie. 2015. *Place in research: Theory, methodology, and methods.* New York: Routledge.

Turner, C. S. V. 2002. Women of color in academe: Living with multiple marginality. *Journal of Higher Education* 73(1):74-93.

Varma, R., and V. Galindo-Sanchez. 2006. Native American women in computing. In *Encyclopedia of gender and information technology* (pp. 914-919). IGI Global.

Walton, G. M. 2020. Questions of Belonging: Arise from Socio-Cultural Contexts Risk Becoming Self-Fulfilling But Can Be Interrupted. Presentation made at the public workshop of the Committee on Addressing the Underrepresentation of Women of Color in Tech, virtual, June 2020. https://www.nationalacademies.org/event/06-04-2020/addressing-the-underrepresentation-of-women-of-color-in-tech-workshop-4-day-2.

Wasburn, M. H. 2004. Appeasing women faculty: A case study in gender politics. *Advancing Women in Leadership Journal* 15.

Washington, G. 2020. Retention of Women of Color Students in Tech: An MSI Perspective. Presentation made at the public workshop of the Committee on Addressing the Underrepresentation of Women of Color in Tech, virtual, April 2020. https://www.nationalacademies.org/event/04-07-2020/addressing-the-underrepresentation-of-women-of-color-in-tech-day-1.

Whitecraft, M. A., and W. M. Williams. 2010. Why aren't more women in computer science. In *Making software: What really works, and why we believe it.* Pp. 221-238. O'Reilly Media, Inc.

Williams, J. C. 2014. Double jeopardy? An empirical study with implications for the debates over implicit bias and intersectionality. *Harvard Journal of Law & Gender*, 37, 185.

Wofford, A. M. 2021. Modeling the pathways to self-confidence for graduate school in computing. *Research in Higher Education* 62(3):359-391.

4

Increasing Recruitment, Retention, and Advancement of Women of Color in the Tech Industry

Recent research findings provide evidence that increasing diversity in the workforce broadens the talent pool, drives innovation and creativity, and increases market growth (Catalyze Tech Working Group, 2021; Hewlett et al., 2013; Morgan Stanley, 2018; Pivotal Ventures and McKinsey & Company, 2018). This chapter describes findings from existing research related to women of color in tech, what is known about their unique challenges in industry settings, and social and environmental factors both inside and outside of tech with the potential to increase their recruitment, retention, and advancement. Lack of comprehensive research data specifically related to women of color in tech was a significant challenge for the committee. As a result, this chapter also draws upon related data on women in STEM professions as well as evidence obtained during the committee's four public information-gathering workshops.

As the number of tech jobs in the U.S. continues to grow, women of color remain a largely untapped pool of intellectual capital and cultural wealth for the tech workforce despite increasing awareness that increased diversity, equity, and inclusion (DEI) can benefit companies (Hodari et al., 2016; Varma, 2018). Women of color make up 39 percent of the female population in the United States and are projected to be the majority of the U.S. female population by 2060. However, women of color are underrepresented in the tech industry relative to their presence in the overall U.S. workforce. Between 2003 and 2019, women made up more than half of the professional workforce (57 percent) but only about one-quarter (26 percent) of the workforce in computing and mathematical occupations, with little improvement since 2007 (DuBow and Gonzalez, 2020; NCWIT, 2020). In the tech workforce, Black women hold 3 percent of jobs, Latinx women hold 1 percent, and Native American/Alaskan Native and Native Hawaiian/Pacific Islander women hold 0.3 percent, and all women of color are underrepresented in leadership positions in the tech workforce (AnitaB.Org, 2019a; Ashcraft et al., 2016). In Silicon Valley, women represent only 16 percent of the Silicon Valley workforce, and women of color from underrepresented groups represent less than 3 percent (Kapor Center, 2018; Pace, 2018).

Although many companies have been increasing their efforts to improve diversity and inclusion for decades, most have not had success in moving the needle. Some companies have been able to show modest improvements in the number of women of color they recruit, retain, and advance, while others have failed to make progress or have even reversed direction (McKinsey & Company, 2020). Women of color in particular remain grossly underrepresented at all levels of the technology workforce, in technology entrepreneurship, and in venture capital investment (Scott et al., 2018).

CHALLENGES

In the sections that follow, the committee discusses barriers to identifying and understanding challenges faced by women of color in the tech industry.

Data Collection and Disaggregation

In the committee's examination of published diversity reports from individual companies and existing research literature on women of color in tech, it found very little comprehensive research literature with data across racial/ethnic groups and gender groups and encountered a lack of intersectional research evidence that included the populations discussed in this report. This lack of robust data poses a significant challenge to clearly identifying the specific challenges that women of color may be experiencing within the tech ecosystem and practices that may be effective; however, existing data do provide some insight into policies and interventions that may help to improve the recruitment, retention, and advancement of women of color. Although U.S. companies are already required to disclose numbers of employees categorized by job category, race, and gender to the Equal Employment Opportunity Commission (EEOC) in their federal EEO-1 reports, they are not required to release this information to the public. Data from voluntary EEOC disclosures shows large racial disparities in the tech workforce relative to the workforce of the private sector overall (Connor, 2017). Over the past few years, more tech companies have begun to publicly disclose their EEO-1 data—driven largely by pledges to improve diversity, equity, and inclusion as well as pressure from investors. However, some hesitation remains due to fears of legal liability, negative publicity, and the potential for data to be used by competitors to attract talented employees to companies with more diversity (Kerber and Jessop, 2020; McGregor, 2020).

Existing available data, prevalent statistical methods, and considerations for protecting the privacy of women of color often obscure the contexts that shape the experiences of women of color. As discussed in previous chapters, in published data, women are often grouped together without disaggregating by race/ethnicity; however, treating women as a homogeneous group hides important insights on outcomes related to racial and ethnic differences and fosters the notion of a universal gender experience despite evidence that the experiences of women of color within the tech industry are not uniform (Shah, 2020). While the lack of disaggregated data related to women of color is a significant challenge in general, some racial/ethnic groups are less likely to be represented in data; Native Americans, Alaskan natives, native Hawaiians, and Pacific Islanders—and the populations within those populations—are often absent from datasets. Like other communities of color, these populations are diverse and could benefit from data-driven, culturally contextualized strategies that take their unique contexts into account.

Existing research on intersectionality has shown that challenges for women of color can be greater than that of racism and sexism combined (Crenshaw, 1991; Kvasney et al., 2009; Malcom et al., 1976; Ong et al., 2011). Intersectional data provide a means by which tech companies could understand systemic patterns and how identity characteristics affect the experiences of women of color (Kvasny et al., 2009). Disaggregation of data is a critical step in understanding the cumulative barriers faced by women of color; the potential strategies for successfully increasing the recruitment, retention, and advancement of women of color as a whole; and targeted strategies that may address the specific inequities faced by smaller subgroups (Catalyze Tech Working Group, 2021). The small number of women of color in the tech sector is often cited as a rationale for grouping all women together; however, disaggregation would provide opportunities to address challenges and improve outcomes across the tech ecosystem. Datasets that are collapsed across race and gender categories in order to use methods of analysis that require more statistical power result in a persistent failure to analyze data related to women of color.

Increasing transparency in data reporting is a critical first step toward creating systems of accountability, understanding the landscape of the tech workforce, and creating opportunities for the tech industry to improve diversity, equity, and inclusion. During its June 2020 workshop, the committee heard from Alice Pawley, associate professor in the School of Engineering Education at Purdue University, on addressing the "small-N" (i.e., small sample size) problem. As a result of the numerical underrepresentation of women of color in tech, their experiences are often understudied and obscured by the practice of combining data for women of color and white women. However, these so-called "small data" can be informative enough to guide industry to actionable solutions and dismantle systems that do not facilitate equity (Metcalf et al., 2018; Pawley, 2020). Alternative methods of analysis that allow for the use of small data, such as the use of qualitative data, can be used to gain insights into how existing systems reinforce inequity and potential strategies for improving diversity, equity, and inclusion. Without this increased transparency in data reporting, accountability will be impossible. Sharing raw, disaggregated employment data publicly will allow companies to measure, standardize, and meaningfully compare data in ways that have the potential not only to highlight trends across the tech sector, but also to drive competition to improve diversity.

Barriers to Entry

Lack of diversity in employment in the tech sector exacerbates the underutilization of available talent as well as the underrecruitment of new talent (EEOC, 2015). For example, EEOC data show that the workforce composition of the top 75 Silicon Valley tech firms is characterized by lack of gender and racial/ethnic diversity (EEOC, 2015, Table 6). Many companies have pointed to a lack of so-called "qualified" candidates in the tech talent pipeline as an explanation for disproportionately low numbers of women of color in tech; however, criteria for what qualifications are necessary and how qualifications are examined are not always well defined. Data show that the rate of people from underrepresented groups graduating with tech degrees is higher than the rate at which they are being hired by tech companies and do not indicate a supply shortage (EEOC, 2015). This section highlights key barriers to entry into the tech sector faced by women of color.

Bias in Recruiting

There is a common belief that disparities are the result of workers leaving the talent pipeline; however, research suggests that the tech pipeline is only one factor impacting the number of women of color in tech. Other factors such as stereotypes and implicit and unconscious bias can drive underutilization of available talent in the tech workforce (EEOC, 2015) as well as result in fewer opportunities to advance, an unwelcoming organizational structure/culture, and underinvestment in diversity, equity, and inclusion efforts (Catalyze Tech Working Group, 2021; Hodari et al., 2016). Lack of diversity in recruitment can often be attributed to a lack of diversity in the places where companies recruit new talent—as they often seek out candidates from elite universities where enrollment for underrepresented groups is much lower (Catalyze Tech Working Group, 2021; Kang and Frankel, 2015; Tiku, 2021). When recruitment departments use college rankings to set priorities for where to recruit and determine resource allocation for recruitment budgets, this gives preference to "elite" universities over universities that produce more graduates from underrepresented groups with tech degrees and works against potential candidates coming to the workforce through alternative pathways. This situation leads to an underinvestment in recruiting candidates of color (Tiku, 2021).

The financial burden required to access some programs created to foster relationships between potential candidates and tech companies may also disproportionately impact women of color who are seeking to enter the tech industry (Nordli, 2019; Tiku, 2021). For example, since 2017, Howard University has partnered with Google to increase access for junior and senior students to industry experience, networking sessions, and opportunities by having them shadow current employees through a three-month program at Google's campus in Mountain View, California. During the pilot phase of this program Howard University and private donors covered the cost of tuition. Pilot funding also paid for housing and a stipend. However, after the pilot phase in 2017, future cohorts became responsible for the cost of tuition, incidentals, and in some cases housing (Tiku, 2021). In 2018, the program expanded to include Hispanic-serving institutions through the Computing Alliance of Hispanic-Serving Institutions. Upon the arrival of the COVID pandemic, Google offered the program virtually; this was able to expand the program's reach. In addition to building technical skills, Google has placed more emphasis on building social capital—building the relationships and networks that can provide people with a sense of identity and belonging (Griffin et al., 2021). This type of program can increase opportunities for internships for students and graduates from historically Black colleges and universities, Hispanic-serving institutions, and other minority-serving institutions that many companies are trying to connect with by equipping students with resources and training that can help them persist in tech. It is critical, however, to consider mechanisms needed to address the economic burden for some that limits accessibility (Kang and Frankel, 2015).

Bias in Hiring

The committee found disparities in the number of women of color earning computing degrees and the number entering the tech workforce. Nationally, Asian women are 32 percent of the female tech workforce, Black women are 7 percent, and Latinx women are 5 percent. In Silicon Valley, these numbers are lower, with Black, Latinx, and Native American/Alaskan Native women in particular representing 2 percent or less of the professional workforce there (McAlear et al., 2018).

In data showing the percentage change in the San Francisco area professional workforce from 2007 to 2015, there was growth in the professional workforce for all groups except Black women. Despite positive growth in the professional workforce for Hispanic women, they still represented less than 2 percent of the professional workforce in 2015. Asian women were shown to be the most likely group of women to be hired into the San Francisco area tech sector (Table 4-1), but the least likely to be promoted (Table 4-2). While Asian women have higher representation in hiring than other women of color, they experience disparities in representation related to advancement in the workforce.

Table 4-1 Change in San Francisco Bay Area Tech Workforce, 2007 to 2015

Cohort	Percentage Change in Professional Workforce from 2007 to 2015	Percentage of Professionals in 2015
White men	31%	32.3%
White women	10%	11.5%
Black men	15%	1.2%
Black women	-13%	0.7%
Hispanic men	32%	3.091%
Hispanic women	11%	1.7%
Asian men	46%	32.3%
Asian women	34%	15.0%

Note: The EEOC defines professional as a job that typically (but not always) requires a professional degree or certification. This category does not include first/mid-level managers or executive/senior level managers.
SOURCE: Adapted from Denise Peck Presentation to the Committee on Addressing the Underrepresentation of Women of Color in Tech (May 14, 2020); Gee and Peck (2017).

These data highlight distinct challenges faced by various groups of women of color. Companies can use these data to better target programs and practices for identifying, interviewing, recruiting, retaining, and advancing women of color.

Barriers to Retention and Advancement

Lack of Training, Support, and Promotion

Across science, tech, and engineering fields, women appear to leave tech at a higher rate—in many cases due to lack of advancement. Among Asian and Hispanic/Latinx women, about one-third report feeling that they are not advancing. For Black women this number jumps to nearly half (Hewlett et al., 2014). Lack of advancement was also cited in a report by the National Center for Women and Information Technology (2020), which found that about one-third of Asian and Hispanic/Latinx women and nearly half of Black women felt stalled in their work. Looking at this from the other side, a 2017 report from AnitaB.Org found that one-third of both male and female senior leaders in science, engineering, and tech companies did not believe that women would reach top positions within their companies (AnitaB.org, 2017). As previously discussed in this chapter, small data sets that are specific to women of color can inform industry's implementation of actionable solutions to address these disparities and can be leveraged to develop best practices.

In 2019, NPower conducted a series of surveys and interviews with alumnae regarding their tech training programs to explore experiences in the workplace, employer inclusion, and other factors that support women of color in tech. Women of color were three times more likely than men to report bias in the form of stereotyping and discrimination, and 24 percent were concerned about gender bias and believed that they were perceived as being less committed or less talented (compared with 1 percent of men). In addition to factors that may cause women of color to leave the tech industry (e.g., lack of advancement and job dissatisfaction), women of color coming from tech training programs may also face challenges that make it harder and unsustainable for them to remain in the work force (e.g., being hired as a contract or part-time worker without benefits) (Shah, 2020; van Nierop, 2020). While training programs may help women of color get in the door and cultivate strategies for success, investing in supports to recruit and retain women of color and providing opportunities for continued professional development and advancement can improve the culture of inclusion (Shah, 2020).

There are many efforts to cultivate high tech skills in women and girls of color both through traditional pathways and alternative pathways (see Chapter 6 for a discussion of alternative pathways). However, evidence suggests that some companies may not be cultivating talent through training of new and existing hires to meet rapidly changing company priorities and needs. This lack of investment in talent development could be a contributing factor to lack of diversity in tech (EEOC, 2015; Hodari et al., 2016; Williams, 2015). Research evidence suggests that systemic barriers within organizations (e.g., lack of mentorship, bias, exclusion from networks of influence, and stereotype bias) can also impede women's advancement to senior-level positions within their organizations (Skervin, 2015).

EEOC data—which are collected for all companies with more than 100 employees and disaggregate by job category/level, race, gender, industry, and geography—are useful when examining trends in recruitment, retention, and advancement among women of color. Figure 4-1 shows promotion trends in the U.S. corporate workforce. Although these data are not specific to the tech industry, they reflect promotion trends that also appear in the tech sector and highlight the importance of reporting employment data with women of color disaggregated from white women to have a clearer picture of the employment landscape.

Denise Peck, executive advisor at Ascend Leadership, presented data to the committee at one of its four information-gathering workshops and provided an example of how these data can be used to create a metric for representation she called the executive parity index (EPI). This metric—the ratio of the percentage representation of a company's executive workforce relative to percentage representation of its entry-level workforce—can be used to examine trends in the number of women of color in the tech workforce over time as well as differences in parity in promotion. Using this calculation, an EPI of greater than 1.0 is above parity (overrepresentation), an EPI of 1.0 is at parity (equal representation), and an EPI of less than 1.0 is below parity (underrepresentation). Looking at data

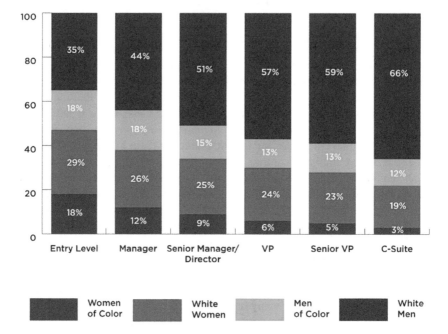

FIGURE 4-1 Representation of the U.S. corporate workforce pipeline by gender and race: Percentage of employees by level at the start of 2020.
SOURCE: Adapted from Lean In and McKinsey, 2020.

representing about 260,000 workers in the San Francisco area tech sector (including Silicon Valley), women were shown to have an EPI of 0.68 while men had an EPI of 1.15 based on 2015 data. However, when the same data for women were disaggregated to compare women by race, the EPI for white women was shown to be 1.17 while the EPIs for Black, Hispanic, and Asian women were 0.61, 0.42, and 0.34, respectively (Table 4-2). Black women were seen to have larger gains in EPI than other women of color during this time frame; however, further analysis attributed this larger gain to a loss of Black women at the entry level rather than a significant increase in the number of Black women at the executive level (Gee and Peck, 2017).

Some have hypothesized that as younger generations—who are assumed to be more adapted to diverse and inclusive workplaces—advance in the tech industry to positions of leadership, there will be a positive effect on diversity and inclusion. While it may still be too early to see effects at the executive level, an examination of advancement of women of color at the managerial level can provide insights as to whether this hypothesis is borne out in current data. Gee and Peck (2017) used the management parity index (MPI)—similar to the EPI—to

Table 4-2 San Francisco Area Executive Parity Index, 2015

Cohort	Executive Parity Index in 2015
White men	72% above parity
White women	17% above parity
Black men	41% below parity
Black women	39% below parity
Hispanic men	11% below parity
Hispanic women	58% below parity
Asian men	38% below parity
Asian women	66% below parity

SOURCE: Adapted from Peck (2020); Gee and Peck (2017).

examine trends in advancement from entry-level professional to middle-level manager (Table 4-3). From 2007 to 2015, the MPI for Hispanic, Black, and Asian tech workers remained relatively constant and was lowest for Asian individuals. This finding does not support the prediction that younger generations of the tech workforce are reducing disparities in advancement. In addition, it indicates that efforts over the past decade to increase diversity at the middle-management level have not had a significant impact on reducing disparity. In this same time frame, women have made gains in MPI, suggesting that efforts to improve gender diversity have had some success. White women had the highest MPI (1.45), and while the MPI for Black women (1.28) and Hispanic women (1.13) was above parity, it increased at a lower rate than MPI for white women; the MPI for Asian women remained below parity (0.69). These data show that reductions in gender gaps were greater than for racial gaps, suggesting that race may play a greater role than gender in preventing advancement (Gee and Peck, 2017).

Table 4-3 San Francisco Area Management Parity Index, 2015

Cohort	Management Parity Index in 2015
White men	25% above parity
White women	45% above parity
Black men	7% above parity
Black women	28% above parity
Hispanic men	32% above parity
Hispanic women	13% above parity
Asian men	31% below parity
Asian women	31% below parity

SOURCE: Data from Gee and Peck (2017).

Attrition

As discussed in Chapter 3, research indicates that structural barriers often lead women to leave the technical workplace: inhospitable work culture/unconscious bias, conflict with preferred work style, isolation, supervisory relationships, promotion processes/lack of advancement, and competing life responsibilities (Ashcraft et al., 2016). A survey by Capital One found that only 2 percent of women in tech left as a result of being unhappy with the nature of their work itself (AnitaB.Org, 2019b). In addition, the National Center for Women and Information Technology estimated that the cost of turnover as a result of bias costs U.S. companies $64 billion dollars a year and that 56 percent of mid-career women in technology are leaving their careers as a result of negative experiences (e.g., institutional barriers and gender bias) in the workplace (Ashcraft et al., 2016).

Williams (2015) suggested that patterns of bias that align with these structural barriers may be a key factor driving women from STEM jobs based on the results of a survey of 557 female scientists and interviews with 60 female scientists:

- Two-thirds of the women surveyed and interviewed reported feeling they had to repeatedly prove themselves and having their expertise called into question. For Black women, this number jumped to three-quarters of respondents.
- Women also felt pressure to balance being seen as too feminine with being seen as too masculine. More than half of respondents reporting feeling backlash from behaving in a way that was perceived as too "masculine," and more than one-third felt pressured to behave in a more feminine way. Being viewed as too feminine often resulted in their competence being questioned, while being viewed as too masculine resulted in them being perceived as aggressive and unlikeable. Black and Latinx women were more likely to be viewed as angry when they deviated from these gender norms.
- Women with children across all racial/ethnic groups felt that their opportunities diminished and their competence was called into question once they had children. They also felt that they had to compete with men who did not have the same level of family/household obligations.
- Seventy-five percent of women felt supported by other women at work, while about 20 percent felt that they had to compete with other women in companies and organizations that were male dominated.
- Isolation was a pattern of bias that seemed to impact Black women and Latinx women in particular. In some cases, this isolation was self-imposed as means of avoiding bias from colleagues—48 percent of Black women and 38 percent of Latinx women reported feeling that engaging with colleagues socially could have a negative impact on perceptions of their

competence and undermine their authority. In other cases, the isolation was perceived as a result of exclusion by colleagues.

Diversity, equity, and inclusion can be critical for companies seeking to retain talent across age groups, and inclusive workplaces may be especially important for early- and mid-career employees. Many organizations struggle to recruit and retain top talent and experience significant turnover. But despite evidence that employees and executives appear to view inclusion as a business imperative and a key factor in determining corporate culture, the inability to implement and operationalize effective diversity, equity, and inclusion strategies may be driving some of this turnover. In a 2017 survey of 1,300 employees across multiple organizations and industries in the United States, Deloitte found that, across all industries, there was a 32 percent increase from 2014 to 2017 in the number of executives who cited inclusion as a top priority—69 percent considered diversity, equity, and inclusion to be an important issue. Eighty percent of employee respondents indicated that inclusion was a key factor in choosing where to work, and 39 percent said they would leave their current employer for a more inclusive one. Among millennials this number was even higher—over 50 percent of millennials would leave their current employer for a more inclusive organization, and nearly one-third of millennials indicated they had already left a less inclusive workplace for a more inclusive one. Millennials were also more focused on inclusion (rather than only increasing demographic diversity) as a part of overall business strategy. Increased inclusion resulted in higher feelings of engagement and empowerment. Evidence suggests that there is a disparity between the level of inclusion that employees prefer in the workplace and the reality of what they experience—possibly indicating that there is a disconnect between company objectives and how those objectives are translated into policies and practices. With higher turnover in younger age groups, tech companies are likely to see younger employees moving to workplaces where there is more inclusion or leaving industry altogether if disparities continue to be pervasive (Deloitte LLP, 2017).

Inhospitable Organizational Culture

The intersection of social identities, such as gender and race, with professional identities, such as job title or level, plays a role in shaping the culture of a work team. The biases associated with these identities, and the impact that has on influence, can also affect how individuals experience the workplace: who is viewed as having earned their position, who has access to resources, how/whether employees are mentored or sponsored, and who is given opportunities to advance (Ibarra et al., 2010; McClain et al., 2021; Skervin, 2015). Organizational culture—more than individual factors—can be a critical driver of women's success as well as their ability to advance (AnitaB.org, 2017). Shared norms, values, and practices define the power dynamics within an organization—both who has influence as well as which types of influence are viewed as effective. Evidence suggests that companies may have more success attracting and retaining women of color in tech by creating organizational cultures in which they are challenged and rewarded for their work, compensated fairly, have adequate peer and mentor support, and are provided with clear pathways for advancement within the organization (AnitaB.Org, 2019a). Inhospitable workplace culture has implications for the success of efforts to recruit, retain, and advance women of color in the tech sector and prevents companies from adequately leveraging the cultures and values of all employees to innovate and develop products and services that mirror the customers they serve (Catalyze Tech Working Group, 2021).

Research suggests that there are four critical aspects of organizational culture across sectors that foster a culture of inclusivity: inclusive leadership, being able to be ones authentic self, access to networking opportunities, and potential for career advancement (Sherbin and Rashid, 2017). Shah (2020) found that among companies with a reputation for gender inclusion there is often support from senior leadership who advocate for inclusive policies and practices and who make clear that pathways to advancement are available. As previously discussed in this chapter, women of color from all backgrounds are less likely to be promoted into positions of leadership in the tech industry where they would be more empowered to implement policies that foster inclusion in leadership.

Evidence also shows that members of groups that are not equitably represented often feel the need to alter their behavior in order to conform to organizational/professional norms in response to prejudiced work environments (Murphy et al., 2018; NASEM, 2021). As previously noted, the talent pipeline for entry into tech careers is sometimes limited by where employers choose to recruit as well as determinations about what type of employee is believed to be the right fit for an organization's culture. This has implications for people coming to tech from backgrounds or pathways that are underrepresented (Catalyze Tech Working Group, 2021). These individuals may feel more pressure to hide parts of their identity to conform or may struggle to find an identity within an environment where they are in the minority, creating an environment that is not designed to help all employees thrive. They may also have fewer opportunities to access broad networks of mentors and sponsors who can help them navigate the workplace (Murphy et al., 2018). Changes in organizational culture can be hindered by biases in recruiting if employees are hired based on how they fit into an existing organizational culture as opposed to what they can add to enhance it (Gee and Peck, 2017).

Although a number of the committee's findings were not specific to women of color, research evidence suggests that improvements to organizational culture that could benefit women of color have the potential to improve satisfaction for all employees (AnitaB.org, 2017; Shook and Sweet, 2019). A 2019 Deloitte report that surveyed employees across 10 industries found that 67 percent of women of color felt they needed to downplay parts of their identity at work in order to fit in. All survey participants worked for companies with diversity and inclusion efforts (Deloitte LLP, 2019a). A 2017 report from Deloitte found a disconnect between employees' desires in terms of organizational culture/structure and that which is being provided by companies. The top three most important cultural aspects named by respondents were an atmosphere where they felt they could be themselves, an atmosphere where they felt they had a purpose and made an impact, and a place where work flexibility (remote work, parental leave, flexible scheduling) was a top priority. Most respondents indicated that they were not looking for organizations where they were surrounded only by people similar to them with similar identities. The most frequently cited reason for leaving (33 percent) was not feeling comfortable being themselves at their organization (Deloitte LLP, 2017; Shah, 2020). Improvements and adjustments to organizational culture (e.g., providing women of color with recognition for leadership, contributions of innovative practices, and sharing of best practices) have the potential to improve group cohesion within organizations, break down barriers among employees, and foster cohesion across diverse groups within companies so that employees from different backgrounds are able to bring their authentic identities to work (Deloitte LLP, 2019a).

PROMISING STRATEGIES AND PRACTICES

Reducing the disparities in recruitment, retention, and advancement experienced by women of color is possible when diversity, equity, and inclusion are elevated as mainstream business imperatives and embedded in organizational culture (McNeely and Fealing, 2018). As a strategic undertaking, diversity, equity, and inclusion need to be resourced, measured, managed, innovated, and rewarded (Box 4-1). This section discusses promising strategies and practices for effecting change within the tech industry.

Prioritization of Diversity, Equity, and Inclusion by Organizational Leadership

Data show that gender and racial/ethnic diversity at the executive level is not enough to achieve improved business outcomes on its own; however, organization leadership is a key factor in determining the success of efforts to improve diversity, equity, and inclusion and can improve business outcomes overall (Catalyze Tech Working Group, 2021; McKinsey & Company, 2018). Companies with more diversity in leadership were also 35 percent more likely to have higher financial returns than the national median in their industry. Companies with diverse leadership teams have been shown to increase profitability relative to peers without diversity in leadership (Morgan Stanley, 2018). A 2015 report by McKinsey &

BOX 4-1
A Snapshot of Lockheed Martin's Approach
to Diversity and Inclusion

In its 2019 *Global Diversity and Inclusion Annual Report*, Lockheed Martin Corporation stated its commitment to diversity and inclusion as a business imperative, explaining that it saw the best way to develop technologies as through a workforce that draws from a diverse set of backgrounds, experiences, and skills. To demonstrate this commitment and underscore the critical role of company leadership in ensuring the success of this endeavor, Lockheed developed an executive inclusion council to approve of diversity and inclusion–related company strategies, composed of senior executive leaders (including its president and chief executive officer, Marillyn Hewson; chief financial officer; and global diversity and inclusion president), representatives from business area functions, and representatives from functional support (Lockheed Martin, 2019).

Lockheed reports diversity and inclusion–focused programs and strategies in the areas of mentorship and networking. In 2011, the company implemented the Program Management and Functional Talent Management initiative, designed to accelerate the development of female and minority talent. The initiative involves intensive talent assessments, accelerated development activities, leadership engagement and accountability, high-visibility opportunities, and exposure to senior executives. The company states that it has seen progress in the number of women and people of color who are nominated to participate, who graduate, and who receive promotions post-graduation (DBP, 2020), although these numbers have not been made public.

In the area of recruitment, Lockheed Martin has developed a strategy to engage minority-serving institutions, particularly historically Black colleges and

---

Company found that for every 10 percent increase in racial/ethnic diversity at the senior-executive level, company earnings increased by 0.8 percent.[1]

All leaders within an organization play a crucial role in creating an inclusive culture (Catalyze Tech Working Group, 2021; Hodari et al., 2016). Deloitte (2017) found that, across all generations, survey respondents preferred leadership that demonstrated inclusive behaviors and said this was a factor in how inclusive they felt their workplace was. Commitment from leadership to embrace diversity, equity, and inclusion as a business imperative can lead to improvements in recruitment and retention of top talent, growth of the company's customer base, and increases in financial returns (Catalyze Tech Working Group, 2021; Kang and Frankel, 2015; McKinsey & Company, 2015). Although not exclusive to the tech sector, a survey by Hewlett and colleagues (2013) with a nationally representative sample of 1,800 professionals, 40 case studies, and numerous focus groups and interviews found that without diverse leadership, women were 20 percent

---

[1] This study was not limited to companies in the U.S. tech sector.

universities and Hispanic-serving institutions (Lockheed Martin, 2019). In 2016 Lockheed Martin launched an initiative with historically Black colleges and universities to provide funding that supports undergraduate research and programs in science, technology, engineering, and mathematics. In their outreach to historically Black colleges and universities, a Lockheed Martin senior leader serves as an executive sponsor. Recruitment efforts at minority-serving institutions have led to a 12 percent increase from 2018 to 2020 in new hires who are women of color (Lockheed Martin, 2016).

To create forums for open and candid discussions around diversity and inclusion among senior leaders, the company launched its Executive Leadership of Inclusive Teams (ELIOT) learning labs in 2006. The learning labs included three sessions—white men's caucus, white men and allies, and ELOIT summits. These sessions were focused on developing partnership skills that deepen diversity dialogues by exposing leaders to concepts of implicit bias and privilege and exploring subordinate group differences and dominant cultures. The sessions were developed based on employee surveys and in partnership with an external consultant. The company solicited feedback on pilot sessions to shape its approach. In 2009, all leadership including vice presidents and above were required to attend one of the three-day ELOIT labs (Lockheed Martin, 2016).

Rania Washington, the former vice president of global diversity and inclusion, has stated that the company continually gathers feedback from employees through surveys to measure the success of its diversity and inclusion efforts. The company utilizes an inclusion index to analyze employee sentiments across various demographics in order to identify successful approaches and areas for improvement; however, details on this index are not publicly available (DBP, 2020).

less likely to have their ideas supported and people of color were 24 percent less likely. In addition, Shah (2020) found that 28 percent of women had concerns about the lack of women role models within their organization.

A crucial strategy for beginning to improve diversity, equity, and inclusion is the development of metrics to measure progress and embedding strategies throughout organizational culture and the daily experiences of employees (Catalyze Tech Working Group, 2021; Deloitte LLP, 2017). This includes defining goals for diversity and inclusion, understanding the resources that are necessary to achieve those goals, and creating structures for long-term accountability for meeting goals. As previously mentioned, metrics that are developed to better understand the engagement and experiences of employees can provide valuable insights into whether strategies to improve diversity, equity, and inclusion are effective (e.g., how employees are doing, whether there is high attrition, and why they are staying or leaving).

In recent years, demand for diversity, equity, and inclusion professionals—and chief diversity officers (CDOs) in particular—has increased substantially across all industries as companies recognize diversity, equity, and inclusion as a priority requiring adequate staffing, funding, and commitment at the highest levels of leadership. Research suggests that there are specific powers and skills that a CDO must have in order to help implement changes that will improve diversity, equity, and inclusion. These include the ability to influence leadership and other contributors at all levels of the business, create metrics to improve accountability, develop business strategies, and achieve concrete business goals (Mallick, 2020). Evidence suggests that the role of CDO is best positioned in senior leadership where they are able to report to (and have the support of) the chief executive officer and are connected to other departments such as human resources, legal, and communications (Catalyze Tech Working Group, 2021; Mallick, 2020; Tiku, 2020). Adequate resourcing of the CDO position includes having a sufficient budget as well as a team to support the work necessary to drive change and allow employees to connect—including training, strategic planning, assessment of the organizational landscape, improving recruitment practices, and building partnerships across industry (DBP, 2009). The implementation of successful strategies may not occur quickly and may need to be iterative, but with clear metrics, companies can better understand their progress.

Improving Data Collection

As previously noted in this chapter, the committee found that most data related to employment in the tech industry were not disaggregated using an intersectional approach to separate out data for women of color. In order to develop metrics and strategies to improve the representation of women of color in tech, disaggregated data are needed that can highlight potential systemic biases; challenges related to recruitment, retention, and advancement; and opportunities to target promising approaches to increase recruitment, retention, and advancement of women of color.

As more tech companies have begun to increase transparency by making their diversity reports and EEOC data public, a clearer picture of diversity, equity, and inclusion in the tech ecosystem is emerging (Box 4-2). During its May 2020 public workshop, the committee heard from Stephanie Lampkin, chief executive officer of Blendoor, a company using data science, demographic statistics, and modeling to analyze existing documentation such as diversity reports, Securities and Exchange Commission filings, company websites, and aggregated employer reviews and ratings to provide diversity, equity, and inclusion insights. These analyses can help companies better understand the race and gender composition of executive teams; overall workplace diversity; diversity of new hires; and the benefits, programs, initiatives, and investments that individual companies are making focused on diversity, equity, and inclusion. Over time, these data allow companies not only to track their progress, but also to compare progress across industries and establish benchmarks. This kind of data also provides industry

---

BOX 4-2
Workforce Diversity, Retention,
and Data Transparency at Intel

In 2015 Intel pledged to increase the company's workforce diversity to achieve full representation of women and underrepresented minorities in its U.S. workforce within five years (Alter, 2015). On its journey to meet this goal the company noted that retention was a key obstacle and implemented two initiatives to address this issue: (1) the WarmLine program, and (2) a comprehensive multicultural retention and career progression study. WarmLine is a service that provides U.S. employees an opportunity to anonymously relay their concerns and issues with a case manager in an effort to resolve workplace struggles before the employees consider leaving the company. Since its launch in 2016, WarmLine has handled more than 20,000 cases, with the top two concerns being lack of career progression and issues with managers (Brown, 2017; Vasel, 2019). The company said in 2018 that it had a 90 percent retention rate among those who had used the program (HPC Wire, 2019). In 2017 Intel undertook a study of retention and career progression, conducting qualitative interviews and surveying over 15,000 employees. The results provided Intel's executive leadership insight into the challenges faced by employees of color. Ultimately, the data from both the WarmLine program and the study were slated to populate tailored playbooks for each of Intel's business units to improve retention, career progression, and representation (Brown, 2017).

Intel has also expressed its commitment to continue to show the company's progress and address its shortfalls in diversity and inclusion efforts by making its workforce data more transparent. While it is not required, Intel has been sharing EEO data since 2003 and publicly disclosed EEO-1 pay data from 2017 to 2019 for all U.S. employees (Intel, 2019). The company has also reported that from 2015 to 2019 the overall representation of women in Intel's U.S. workforce increased by 6.9 percent, although it declined slightly from 26.8 percent in 2018 to 26.5 percent in 2019. Meanwhile, the overall representation of employees from underrepresented groups increased in the same four-year period to 15.8 percent (Estrada, 2020). In its 2019 diversity reporting Intel introduced new data for women of color, noting that they make up 3.8 percent of its U.S. workforce.

While Intel does not further parse its data into individual categories of race and ethnicity publicly, Dawn Jones, Intel's chief diversity and inclusion officer, has expressed hope that looking at this kind of intersectional data will help Intel recruit and retain women of color, particularly in leadership roles (Carson, 2020). Looking forward, the company has pledged to put women in 40 percent of technical roles globally and to double the numbers of women and people from underrepresented groups in leadership roles in the next 10 years (Intel, 2020).

leaders with information necessary to understand the scope of the challenge, develop targeted solutions, and improve their outcomes (AnitaB.org, 2017; Catalyze Tech Working Group, 2021; Lampkin, 2020).

Recruiting, Supporting, and Developing Talent

Attracting women to the tech industry requires implementation of practices that foster success and provide opportunities for their professional development. As previously discussed in this chapter, women of color face a number of challenges and systemic barriers as they enter the workforce; however, many of these challenges that negatively impact their experiences (e.g., discrimination, salary disparities, isolation, imposter syndrome, inhospitable organizational culture) persist even as women of color continue to advance in their careers. The committee discusses promising practices for recruiting, supporting, and developing talent at all levels in the sections that follow.

Recruiting and Hiring

Companies can adopt strategies that leverage existing mechanisms for recruiting women of color in order to build organizations that are inclusive from end to end and attract more diverse candidate pools that reflect the customers that businesses serve (Mallick, 2020). Some examples include training diverse recruiting teams in using best practices for interviewing and reducing bias in hiring, having recruiters to partner with professional organizations for women of color (see Chapter 6 for a discussion of alternative pathways and the role professional organizations), recruiting at conferences for women of color, reducing the emphasis on culture fit in the hiring process, using language in job advertisements that avoids more masculine-gendered words (e.g., competitive and assertive), utilizing recruitment tools and software to proactively seek out women of color as potential employees, and giving hiring managers blind resumes.

Women of color who are new to technology or corporate America may not have access to the same types of resources from peers, parents, or friends who help them navigate their roles in industry; there is therefore a need for resources to support them (Catalyze Tech Working Group, 2021; Shah, 2020). For example, in internship programs, there is often not recognition of the differing needs that interns may have, which is needed to support them during the internship and potentially bring them into the company in the future. For interns who are women of color this could be support for dealing with bias in the workplace, and mentors who can provide guidance and coaching to help them navigate the interview process and other challenges they may face in the workplace.

Partnering with higher education institutions that are successful at retaining women of color in technology and computing majors can also benefit companies by creating opportunities to share best practices for supporting women of color.

At its April 2020 workshop, the committee heard from Gloria Washington, assistant professor in the Department of Electrical Engineering and Computer Science at Howard University, who highlighted the importance of symbiotic relationships when building partnerships between industry and higher education institutions. For example, while minority-serving institutions may benefit from having an industry presence on campus (e.g., as guest lecturers or during recruitment efforts), it is important that tech companies understand the unique environment and culture of these institutions and how they allow students to flourish. Institutions of higher education can be a valuable partner to industry by helping to show tech companies how to improve mentorship experiences for students and interns, and increase the opportunities to provide students with exposure to the entrepreneurial spirit of the tech industry (Washington, 2020).

Retention and Advancement

Evidence suggests that increasing the transparency of steps needed to advance in a company can reduce bias in the promotion process. In particular, promotion processes that are well defined with incremental goals can help clarify steps for employees who are seeking to advance as well as give managers a clear understanding of which metrics should be used for promotion and allow them to link advancement to measured performance (AnitaB.org, 2017). Linking advancement to performance provides more opportunities for recognition and can improve employee engagement. In addition, having metrics that track business processes improves data collection by providing companies with additional information that can be measured over time to understand whether strategies and approaches for reducing bias and improving retention and advancement are successful.

One promising model for changing business systems to eliminate bias is an approach developed by Joan Williams: metrics-driven bias interrupters (Williams et al., 2016). These strategies focus on basic business practices and are intended to reduce bias through objective, iterative, and scalable redesign, such as modifying job descriptions in small but significant ways, ensuring that there is equity in high-profile work assignments, defining clear criteria for promotion, and seeking signs of bias in performance evaluations. The bias interrupter model has the following four basic steps:

- Assess: investigate whether and how biases are operating within business processes and identify objective metrics that can measure whether bias exists.
- Implement a bias interrupter: employ an intervention to interrupt biases that have been identified.
- Experiment and measure: determine whether the intervention interrupted bias sufficiently and whether metrics improved.

- Increase interrupter if necessary: increase or modify the interrupter if metrics have not shown sufficient improvement.

Increasing opportunities to equip women of color with coaching and training to help them navigate the tech ecosystem may also increase the likelihood that they persist in tech. Rati Thanawala, founder of the Leadership Academy for Women of Color in Tech—a program under development—presented at the committee's February 2020 workshop. As part of the development of the academy, Thanawala was exploring the experiences of successful women of color in tech to help identify effective strategies and levers of success. Using these strategies, women of color enrolled in the leadership academy, which would begin in graduate school and continue for three to five years as the women transition to jobs in industry, would receive long-term coaching on strategies for success (e.g., increasing leadership training, developing negotiating skills, understanding career pacing, and raising their professional profile). Thanawala also discussed how partnerships between colleges and universities and the tech industry as well as technology leadership councils help make this type of coaching more widely available by investing in it for women of color. Box 4-3 highlights an example of efforts to increase the number of women in senior and executive roles at IBM.

Employee Resource Groups

Evidence suggests that women of color benefit from peer networks and "counterspaces" (i.e., safe spaces) at educational institutions, in the workplace, within professional societies, and in relationships with mentors, sponsors, and colleagues. Counterspaces are particularly important because they serve as outlets where women of color can vent frustrations, share and validate their experiences, and build positive racial identities (Ong et al., 2018; Solórzano and Tara, 2003). Employee resource groups—sometimes called employee affinity groups—are one example of a counterspace that can play an important role in helping companies to foster a more diverse and inclusive workplace, encourage professional development, and increase a sense of belonging for employees across an organization. Participation in an employee resource group can also provide women of color and other employees with more access and visibility to senior management (Catalyze Tech Working Group, 2021; Tiku, 2020).

In order for employee resource groups to effect change within organizations, evidence suggests that they should be empowered to push for changes without fear of retribution and should be a complement to rather than a replacement for diversity, equity, and inclusion professionals in leadership positions. Evidence suggests that in some cases, members of employee resource groups are called upon to take on additional responsibilities related to diversity efforts (e.g., serving as brand ambassadors, helping to recruit new employees, and serving on focus groups for policies) (Tiku, 2020); however, these additional responsibilities

---

BOX 4-3
Advancing Women with Leadership Potential at IBM

In its 2020 diversity and inclusion reporting IBM called attention to its investments to increase the representation of women employees in senior and executive roles. IBM has stated that the inclusion of women on senior teams has had a positive effect on company performance (IBM, 2017). One program implemented as part of the company's effort to build this imperative into its talent development strategy is Pathways to Technical Leadership. This program assists women in mid-level technical jobs identify career aspirations and develop leadership skills through workshops, mentoring, and training. Senior leaders engage with the program by offering opportunities to shadow other employees and hosting roundtable discussions (AnitaB.org, 2018; IBM, 2017).

Additionally, IBM has emphasized its "skills over degrees" approach, which has opened a route to tech for women with backgrounds outside of tech or who have been absent from the tech field for a significant period of time. One program is IBM Tech Re-Entry, a six-month paid "returnship" for technical professionals who have been out of the workforce to help them rejoin the tech industry (IBM, 2020, p. 40).* Women make up 99 percent of the program's participants. The IBM Apprenticeship Program is a paid apprenticeship that offers participants a chance to build technical skills through "an intensive work-based development program, with comprehensive learning, focused hands-on training, and mentorship" (IBM, 2020, p. 40). IBM has found that this program has created an opportunity for women who do not have four-year degrees to shift from food service, retail, and manufacturing jobs to positions as cybersecurity analysts, system administrators, digital designers, and developers.

While these programs' explicit focus is on all women in general, IBM cites these programs also as pathways to bring more women from diverse backgrounds into the company.

---

*For more information see https://www.ibm.com/employment/techreentry.

---

are not generally part of an employee's official job and can create unintended, additional burdens for employees of color. There are strategies for utilizing the strength and experiences of these groups in a way that increases their benefit to members, such as tying leadership of a group to performance review metrics or providing time to employees to participate in one.

Organizational culture also plays a role in the success of employee resource groups (ERGs). They are most successful when there is buy-in from leadership and engagement from other employees. In recent years, some companies have begun to provide additional compensation to leaders of ERGs for their service as a way of recognizing the value they bring to the organization—both in terms of helping foster a culture of belonging as well as helping companies develop deeper insights about the customers they serve (Montañez, 2021; Morris, 2021). When they are treated as an asset to the business—like other business priorities—these groups can play an important role in helping companies reach organizational goals such as improved recruitment and retention, brand enhancement, training, and professional development (DBP, 2018).

During its May public workshop, the committee heard from Melonie Parker, chief diversity officer at Google, and Bo Young Lee, chief diversity and inclusion officer at Uber. Both presentations provided useful examples of how employee resource groups can help drive organizational change and be an asset to companies. Parker described Google's 16 employee resource groups, which have 35,000 employees who actively participate in the groups. As part of the onboarding process, these groups include a buddy program to help orient new hires to Google. The groups also assist with product design and help provide insights into community sentiment (Parker, 2020). Lee described Uber's 12 employee resource groups as being fundamental to its diversity and inclusion efforts. In 2015, two of the groups—one for people of color and one for women—played a key role in raising the need for diversity and inclusion practices at Uber. As Uber has implemented systemic changes aimed at improving diversity and inclusion over the past five years, Lee noted that employee resource groups have continued to be central to the process. Uber's executive leadership team meets quarterly not only with the chief diversity and inclusion officer, but also with a council of employee resource groups that provide direct feedback to the leadership team (Lee, 2020).

Lockheed Martin also provides a useful case study. The company has employee resource groups and networks, which are voluntary, employee-led groups formed based on various dimensions of diversity including race, ability/disability, ethnicity, gender, gender identity, military/veteran status, age, and sexual orientation. In 2016 the company had 70 employee resource groups encompassing approximately 8,000 members across all business areas. The company considers that these groups provide benefits to both the employees and its business through leadership development, learning and cultural awareness, and increased employee engagement (Lockheed Martin, 2016).

Allyship

Women of color who have strong allies are more likely to feel that they can be their authentic selves at work, have higher job satisfaction, and believe that they have opportunities to advance. Allyship, as defined by Nicole Asong Nfonoyim-Hara, director of diversity programs at the Mayo Clinic, is "when a person of privilege works in solidarity and partnership with a marginalized group of people to help take down the systems that challenge that group's basic rights, equal access, and ability to thrive in our society" (Dickenson, 2021). While most people who are not from underrepresented groups may believe they are an ally of women or people of color, in reality allyship requires intention and a distinct set of behaviors that have to be learned. In 2019, Deloitte surveyed 3,000 nationally representative U.S. adults employed full time at companies of 1,000 or more on workplace inclusion and the impact of bias. Results found that most surveyed individuals (92 percent) believed themselves to be allies, yet few addressed bias when they witnessed or experienced it (29 percent) and nearly one-third ignored it (Deloitte LLP, 2019b).

Deloitte's survey (2019a) also found that 60 percent of respondents felt that bias was present in their workplace, and 83 percent of workers who experienced bias characterized the bias as indirect microaggressions. The top three forms of bias reported were age, gender, and race/ethnicity. Respondents indicated that their performance was negatively impacted by bias even when they were not the one who personally experienced it. Ninety-two percent of respondents in the Deloitte survey felt that they were allies in the workplace, and 73 percent felt comfortable talking about bias in the workplace, yet one-third reported that they ignored bias that they witnessed or experienced. Although not specific to the tech sector, findings from Lean In and McKinsey & Company show similar survey results and also highlight differences in levels of allyship among senior-level men and senior-level women (Figures 4-2 and 4-3).

Bias presents a challenge for organizations that have not developed strategies for addressing bias and increasing allyship, but understanding the intersection of identities increases opportunities for teams to connect with each other and find common ground. Fostering and resourcing practices that create an inclusive

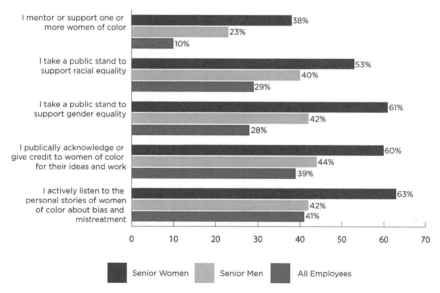

FIGURE 4-2 Senior-level women are much more likely than senior-level men to practice allyship. Figure shows the percentage of men in senior leadership vs. women in senior leadership who consistently take allyship actions.
SOURCE: Adapted from Lean In and McKinsey, 2020.

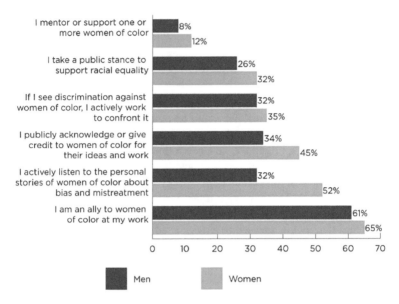

FIGURE 4-3 Employees who say they are allies do not always take action. This figure shows the percentage of employees who say they are allies to women of color vs. percentage who consistently take allyship actions.
SOURCE: Adapted from Lean In and McKinsey, 2020.

organizational culture is a strategy for reducing bias and allows all employees to demonstrate their value. Evidence suggests that the majority of workers want to be supportive of colleagues and already view themselves as supportive, so the "buy-in" is more a question of using effective strategies than it is about literal buy-in. Efforts to decrease bias improve productivity, morale, well-being, and confidence and can decrease turnover. Organizations should take advantage of this tendency toward allyship to further advance inclusion.

Mentorship

Although women of color can benefit from having effective mentors from many different backgrounds, culturally responsive mentorship prior to and during employment can help women of color establish their identity within tech and navigate pathways both into the tech sector and through levels of advancement within companies (Skervin, 2015). A 2019 National Academies report defined mentorship as "a professional, working alliance in which individuals work together over time to support the personal and professional growth, development, and success of the relational partners through the provision of career and psychosocial support" (NASEM, 2019). Although efforts to mentor women of color are generally well intentioned, Marisela Martinez-Cola, assistant professor of sociology at Utah State University and a presenter at the committee's June 2020 workshop, provided the committee with examples of how the attempts at allyship by mentors can often be problematic. Underrepresentation of women of color in the tech industry results in the availability of fewer women with shared, first-hand experiences to serve in this role.

Martinez-Cola described three types of mentors: collectors, nightlights, and allies. "Collectors" may mentor many mentees of color but often limit their interactions to activities and events related to diversity initiatives. They may be well intentioned and may still play a valuable role (e.g., by helping connect mentees with resources), but they may also be inadvertently condescending, negatively affecting the interpersonal dynamic between the mentor and mentee. "Nightlights" have a better understanding of the challenges of people of color, acknowledge systemic racism, and are able to help people of color navigate predominantly white spaces. They also use their privilege and cultural and social capital to make mentees more visible. For example, they may intervene if a woman of color is tokenized or treated as a representative for all women of color. They might also nominate women of color for tasks or roles that are not related to race or difference. Lastly, "allies" are aware of the experiences of their mentees of color and can make meaningful connections in addition to those based on shared lived experiences. They help mentees make connections with important individuals in their field, help them gain standing, and can take criticism and feedback constructively and without damaging the relationship. Mentors who are allies build trust with consistency and humility that affirms their mentee's belonging (Martinez-Cola, 2020).

The NASEM (2016) report found that the success of mentees who did not have access to culturally responsive mentoring experiences could be impeded as they worked to balance and navigate multiple intersecting identities (e.g., race, culture, gender, technical expertise). This conflict can result in depression, reduced psychological well-being, and lower professional performance. Research suggests that recognition of these intersecting individual identities can influence career development and were a key component of effective mentorship. NASEM (2019) also recommended using inclusive approaches that intentionally consider how culture-based dynamics such as imposter syndrome have the potential to negatively influence mentoring relationships, and understanding how biases and prejudices may affect the relationship between mentors and mentees, specifically for mentorship of individuals from underrepresented groups.

In general, women are less likely than men to have access to informal networks in the workplace that can connect them with opportunities to advance and take on high-profile assignments. Mentorship can position women of color for success in the tech industry, but companies can help improve these outcomes by creating accountability structures to ensure that employees have the support needed to succeed such as formalized mentorship programs that set clear expectations and define measurable goals (e.g., expected time commitment, identification of outcomes that will be evaluated) for both mentors and mentees and documentation of growth and performance. Research also suggests that it may be beneficial to have access to more than one mentor so that support can be targeted to different types of challenges (Shah, 2020; Skervin, 2015).

Sponsorship

Research suggests that many companies are not cultivating pathways and opportunities for women of color to advance early in their careers and that 56 percent of women in technical professions leave their positions by mid-career. (AnitaB.org, 2017; Hewlett et al., 2008). Formal leadership development with the support of an executive sponsor is a means by which companies improve advancement and growth by improving retention, reducing isolation, and increasing social capital for women of color (AnitaB.org, 2017). The role of a sponsor differs from that of a mentor. Sponsors are advocates who can help fellow employees increase the visibility of their work—both within and outside of their company—by helping them navigate networks and build relationships. A sponsor should be someone who is senior with the ability not only to support a woman of color, but also advocate for access to opportunities to grow and advance. As with mentorship, accountability plays an important role in determining the effectiveness of sponsorship. Evaluation of whether the employee who is being sponsored advances and has opportunities to serve in visible leadership roles is one way in which a sponsor's success could be evaluated. Intentional strategies for increasing the visibility of women of color is beneficial to companies and increases opportunities for women of color to thrive and advance (AnitaB.org, 2017; Martinez-Cola, 2020; Shah, 2020).

Many women are overmentored and undersponsored. Sponsorship is an especially important part of improving an overall culture of inclusion for women of color. Companies can increase the number of sponsors by training managers on effective strategies for sponsorship. Although not all sponsors of women of color need to be women of color themselves, sponsors should share the core values and goals. For smaller companies, which may have fewer resources or opportunities for advancement, it can be beneficial to connect women with resources in the company as well as opportunities outside the organization (Shah, 2020).

Organizational Culture and Work-Life Balance

In previous discussions of recruitment, retention, and advancement in this chapter, issues related to organizational culture and work-life balance have been highlighted as factors affecting whether women of color persist in tech careers. Intersectionality theory as well as diversity and inclusion concepts are relevant to research on organizational culture and work-life issues. Many organizations are challenged by the long-term commitment to data collection and sharing required for diversity and inclusion work to be successful and to create sustainable changes in organizational culture.

The committee heard examples from industry at its May 2020 workshop about how data and transparency can help to build a more inclusive workplace. Google's Chief Diversity Officer, Melonie Parker, described how Google has been working to increase transparency in its data to understand the root of organizational challenges related to diversity and increase accountability of leadership. Parker's presentation provided a number of examples of steps Google is taking such as

- Looking at data for populations that have historically been denied access or opportunities in the workplace,
- Using an intersectional approach with a focus on equity to understand data and ensure the data are informed by the lived experiences of those groups to avoid creating monolithic practices, and
- Holding summits for Black, Latinx, Indigenous, and Asian women to build community and better understand the needs of each group.

By understanding the experiences of employees, Parker noted that Google aims to tap into a broader potential talent pool, which will help them meet industry demand in coming years (Parker, 2020).

Uber's Chief Diversity and Inclusion Officer, Bo Young Lee, discussed one of Uber's key organizational changes over the past five years—a commitment to making systemic change within the company to improve strategies for hiring, developing, and promoting talent and to create a more inclusive organizational culture. Diversity and inclusion strategies are guided by key principles. Lee described how cultivating empathy, for example, through storytelling and by shifting language (e.g., using the term people of color vs. minority) is one strategy that Uber has used. A second strategy is to increase transparency and accountability. Uber has invested heavily in data collection and making more data public. Beginning in 2019, the company began enhancing collection of employee demographic data to include eight dimensions of diversity (race, ethnicity, gender, sexual orientation, gender identity, veteran status, gender status, and caregiver status), plus a proxy question related to economic status during childhood. These dimensions of diversity help the company ensure that the diversity that is found across its enterprise also exists within its corporate ecosystem. In addition, Uber publishes intersectional data in its annual diversity report, which is publicly available. Access to these demographic data is helping to drive accountability on individual teams as well as for the company as a whole.

The third principle is to invest in long-term change. Lee discussed with the committee how sustained change requires a long-term commitment to implementing effective strategies consistently over time. In Uber's case, she expected sustained change to take at least four to five years; however, implementing these changes creates an environment that is designed for all employees to thrive and where more employees will be retained (Lee, 2020).

It is important to note that while the examples of data collection and data sharing provided in the presentations heard by the committee are commendable, increased data availability and transparency are only one part of increasing diversity, equity, and inclusion within an organization. Organizational culture and leadership are key drivers of how data are utilized to effect meaningful organizational change. There have been several recent examples of alleged discrimination within a number tech companies—including those who presented to the committee—that highlight challenges that women of color continue to face in corporate tech culture. Allegations of gender and race discrimination in performance evaluation and promotion (Bort, 2020; Duffy, 2020; Reuters, 2018), discrimination in recruitment and hiring (Duffy, 2020; Dwoskin, 2020), racist and misogynistic harassment (Bort, 2020), retaliation (Duffy, 2020), and termination of employment (Allyn, 2020; Duffy, 2020; Gupta and Tulshyan, 2021; Simonite, 2021) have been reported in the news and have prompted congressional inquiries and investigation by the Equal Employment Opportunity Commission (Bond, 2021; Dwoskin, 2020). The persistence of these types of issues discourages women of color from pursuing or remaining in careers in tech and highlights the progress yet to be made in creating inclusive workplaces. In addition to increasing transparency around the number and demographic makeup of women of color in tech, there is an opportunity to better understand how organizational culture shapes their experiences in the workplace. The experiences of women of color can provide critical insights and inform the development of effective programs and practices as companies work to foster a diverse, inclusive, and equitable culture and long-term organizational transformation.

Research focused on work-life issues in an academic context suggests that work-life issues related to gender have often been prioritized over issues related to race, especially as those issues intersect with gender (NASEM, 2021); however, understanding the work-life preferences of women of color is an important part of understanding how to ensure that both their personal and professional needs are met. Evidence shows that individuals who work in environments where both the organizational culture and leadership support work-life balance are less likely to experience conflict between their work and home life and to view their employer as supportive of work-life balance (NASEM, 2021). The expectation to be available to work at all hours, a lack of flexibility work around family care obligations, and expectations for extensive work travel, for example, are factors that can negatively affect the ability of women of color to advance, feel a sense of job security, or remain with their current employer (Metcalf et al., 2018; Murphy et al., 2018; Skervin, 2015). In a 2016 evaluation of top companies for women technologists focused on hiring, retention, and advancement of women in technical roles, flex-time policies were a key differentiator for companies with above average representation of women in these roles. Eighty-eight percent had opportunities for remote work, 83 percent offered flexible hours (i.e., which hours in the day to work), and 79 percent offered flexible schedules (i.e., number of days per week) (AnitaB.org, 2017). The COVID-19 pandemic has highlighted the triple burden that many women share—childcare, eldercare, and work (Box 4-4).

Increasing Industry Collaboration

The scope of disparities in the tech workforce became more apparent in 2014 following the disclosure of EEOC data by a number of top tech companies. Examination of disclosures since 2014 shows trends across the tech sector that are consistent with findings at individual companies (Gee and Peck, 2017). The seemingly universal nature of these trends highlights an opportunity for intra- and cross-sector collaboration to address the disparities. Some companies have already begun to work as part of collectives where they can share best practices, aggregate and analyze intersectional data, leverage the insights of industry leaders and employees, and develop metrics to improve diversity, equity, and inclusion in the tech industry (Gee and Peck, 2017).

Investing in Women of Color

There is substantial underrepresentation of women of color in entrepreneur-ship as well as in venture-invested firms despite the fact that women of color make up 80 percent of women creating small businesses (Kapor Center, 2018). In addition, only 1 percent of venture capitalists are Black women and less than 1 percent are Hispanic women (Huaranca Mendoza, 2020; Kapor Center, 2018). Although limited intersectional data exist on entrepreneurship by women of color, evidence suggests that the number of start-ups founded by women of color is low (McAlear et al., 2018). Similar to the dynamic observed in recruitment in the tech sector, venture capital firms—primarily led by white men—often fund start-ups that they connect to through existing business contacts, the majority of whom are also white men. Carolina Huaranca Mendoza, founder of 1504 Ventures, presented to the committee at its May 2020 workshop and highlighted how difficult it is for people of color to create their own venture funds because of the need to already have sufficient personal wealth to invest, which people of color are less likely to have (Huaranca Mendoza, 2020). Factors such as these result in a repeating pattern of disproportionate exclusion for entrepreneurs of color (Kang and Frankel, 2015; Morgan Stanley, 2018; Women Who Tech, 2020).

Morgan Stanley conducted a survey of 101 investors and 168 bank loan officers to examine the funding gap for female and multicultural entrepreneurs and found a disparity between investors' perceptions of how much they were investing in women- and minority-owned businesses and how much they were actually investing. Most investors believed their investments to be more equally

BOX 4-4
The Impact of COVID-19 on Women of Color in Tech

Organizational policies, job structures, and industry norms can often dictate the extent to which individuals are able to manage their workload and balance family and job obligations and maintain boundaries between work and home. A recent National Academies study highlights research showing that permeable boundaries between work and home can increase employee distress, increase turnover, and negatively impact work performance (NASEM, 2021). Although workers whose jobs could be performed remotely were less likely to be affected, corporate America has been disrupted in unanticipated ways across sectors as a result of COVID-19, and women have faced distinct challenges with one in four women considering reducing their work hours or leaving the workforce completely (Lean In and McKinsey, 2020; NASEM, 2021). Research suggests that there are differences in work-family demands faced by women and men, yet despite improvements in division of family responsibilities between men and women, inequities in work-life balance persist. These inequities have become especially pernicious as companies struggle to pivot and adapt to the COVID crisis and the new version of normal it has created.

Communities of color have experienced disproportionate disruption to employment during the pandemic (Brown, 2020). For example, a survey by Project Include, a non-profit focused on advancing data-driven DEI solutions, found that tech workers reported increased rates of harassment related to race, gender, and age since the start of the pandemic (Catalyze Tech Working Group, 2021). A recent McKinsey & Company report (2020) found that continuing diversity and inclusion efforts during the pandemic (rather than deprioritizing them) was beneficial to companies. In companies where diversity and inclusion are deprioritized, women, people from underrepresented groups, and women of color are likely to have their jobs disproportionately put at risk by an economic downturn even when their jobs can be performed from home. In addition to being more likely to have more responsibility to provide care to children or family members and to supervise children's remote schooling, women of color may be less likely to have access to adequate workspace for remote work at home (e.g., adequate broadband access,

---

distributed than they were in reality (Morgan Stanley, 2018). A survey by Women Who Tech found that 70 percent of investors believed that the size of the pipeline of potential companies to invest in was the cause of disparities, and 56 percent did not believe that access to funding was inequitable (Women Who Tech, 2020). However, in 2020, the median seed round funding for Black women was $125,000 and $200,000 for Latinx women, compared with a national median of $2.5 million dollars (Kapin, 2020). This disconnect between perception and reality plays a role in perpetuating imbalances in investments in women of color, leading to underinvestment in their businesses.

equipment provided by an employer), putting them at a disadvantage compared to other employees.

Several speakers at the committee's public workshops in spring 2020 highlighted many of the challenges that specific communities of women of color have faced during the pandemic. Rocío van Nierop, co-founder and executive director of Latinas in Tech, highlighted one of the ways that the pandemic affected the Latinx workforce. Many companies have increased workforce diversity by creating positions for contingent workers rather than full employees; however, these positions were often the first positions to be eliminated as companies moved to reduce costs as a result of COVID-19. The loss of these positions disproportionately affected women of color (van Nierop, 2020). The committee also heard about the infrastructure challenges affecting Native American communities from Andrea Delgado-Olson, founder and chair of Native American Women in Computing. Lack of internet connectivity was exacerbated in communities that were already experiencing a significant digital divide. Data from 2018 reported that 27 percent of Native American households were without internet access or had only dial-up access. For individuals living on reservations where internet access is limited or does not exist, remote working and schooling might require driving miles off reservation to public libraries or other establishments with open wireless connections in order to work or complete school work (Delgado-Olsen, 2020). Many of these establishments were forced to close to the public during the pandemic, increasing the challenge of staying connected.

McKinsey & Company (2020) suggested that diverse and inclusive companies may be better positioned to adapt to the COVID-19 crisis and other future crises that cause workforce disruption, because they are more likely to make bold decisions that help them to shift and innovate in order to address challenges and develop solutions that allow employees to continue their work, rather than losing talent to downsizing. The pandemic has already highlighted ways in which virtual settings may help to increase equitable access to opportunities—for example, creation of virtual internships that allowed students to stay closer to their communities and families. The COVID pandemic presents an opportunity for companies to reimagine how they utilize remote work and other flexible work policies, and make investments in infrastructure in a post-pandemic future to increase access to a more diverse workforce and a more equitable future for all employees.

Bias can also play a role in influencing investor decision making related to women- and minority-owned businesses. Responses to the Morgan Stanley (2018) survey indicated that investors were more likely to have a preconception that these businesses perform below market average and are riskier investments. Investors were also less likely to understand ideas for businesses whose products and services targeted customer bases they did not relate to—groups that women and multicultural entrepreneurs are more likely to serve. Investors who were women or from underrepresented groups were more likely than white male investors to review opportunities from women- and minority-owned businesses. The use of intersectional data would elucidate how these biases compound to impact the success of women of color in acquiring investment.

However, a promising finding from the report shows that the key to achieving greater parity in successfully obtaining investment may be increasing opportunities for women- and minority-owned businesses to meet with potential investors (Morgan Stanley, 2018). Support for women of color as they prepare to take their ideas to investors can include coaching, creating communities of entrepreneurs, and fostering mentor networks. Venture capital companies can also increase racial and gender equity by making themselves more accessible to these communities of entrepreneurs and providing investments in later rounds of funding. In addition, foundations that invest in venture capital funds can help increase the numbers of entrepreneurs who are women of color by creating mandates to identify general partners who are women of color, cultivating emerging managers who are women of color, investing in mission-driven for-profit companies, and investing in programming that targets women of color (Huaranca Mendoza, 2020). Parity in venture investment will allow businesses and start-ups owned by women of color to increase innovation and maximize potential returns in ways that are mutually beneficial to entrepreneurs and investors alike (Catalyze Tech Working Group, 2021; Pivotal Ventures and McKinsey & Company, 2018; Shah, 2020). Disparities in investments in women- and minority-owned businesses are also present in the distribution of public-sector grants such as the U.S. government's Small Business Innovation Research and Small Business Technology Transfer programs, which are intended to fund research, development, and commercialization of technological innovation by small businesses. These programs award approximately $3.5 billion dollars per year in funding to small businesses and are mandated to support participation of women, individuals who are socially or economically disadvantaged,[2] and businesses from underrepresented areas. Data reporting the distribution of awards between 2005 and 2017 showed disparities in support for women- and minority-owned businesses. Awards for women rose from 8 percent to 11 percent while awards to socially or economically disadvantaged business owners remained at 8 percent over that time span (Liu and Parilla, 2019). Data disaggregation by race/ethnicity and gender would be beneficial to further understand the scope of disparities in investments in women of color.

Corporate philanthropy can play a role in improving the representation of women of color in tech. However, companies often struggle to know how to make this kind of high-impact philanthropic investment and often rely on self-guided internet searches of research to guide their giving (McKinsey & Company, 2020). But some efforts to coordinate these philanthropic efforts across the tech sector are under way (Box 4-5).

---

[2] This is defined as a small business that is 51 percent or more owned by one or more persons from any of the following groups: Black Americans, Hispanic Americans, Native American, Asian-Pacific Americans, Asian Americans, and other groups as determined by the Small Business Administration.

BOX 4-5
Reboot Representation

During its May 2020 workshop, the committee heard from Renee Wittemeyer, senior lead for tech innovation at Pivotal Ventures, an investment and incubation company working to advance social progress, and Dwana Franklin-Davis, chief executive officer of Reboot Representation. In 2018, Pivotal Ventures led a study drawing on insights from 32 leading tech companies as well as 100 executives and sector experts to understand how and whether tech companies were using their own profit dollars to close the gender gap in tech through corporate philanthropy and corporate social responsibility giving, and to understand how much they were spending. Despite companies' expressing a strong desire to close the gender gap in tech, only 5 percent of the surveyed companies' philanthropic giving actually went to programs with an explicit focus on women and girls, with less than 0.1 percent going specifically to women of color in tech. Pivotal was able to use this report to help provide companies with toolkits and templates to better evaluate the impact of their giving and to coordinate efforts between human resources and corporate responsibility/philanthropy efforts. The report catalyzed a first-of-its-kind coalition of 15 tech companies—called Reboot Representation—that pooled over $17 million toward the goal of grantmaking to double the number of Black, Latinx, and Native American women graduating with computing degrees by 2025.

These companies are using their collective resources to try to solve an industry-wide problem by pooling philanthropic dollars to make strategic investments in computer science education for Black, Latinx, and Native women; raise the profile of the issue of underrepresentation through communication activities; and bring together companies to share best practices and lessons learned with a unified voice. Although the mission of Reboot Representation focuses on Black, Latinx, and Native women, the hope is that the practices that improve conditions for the populations of focus will inform strategies that benefit all women of color. Franklin-Davis underscored the importance of aligning corporate-giving goals with corporations' organizational goals, diversifying the pool of universities from which companies recruit, and working with universities to make sure that there is alignment between their educational programs and industry's needs. She highlighted four areas where industry could make investments: (1) disaggregation of data to improve strategic planning and targeted grant investment; (2) investments in early exposure to computer science for women and girls of color to provide them with computer science fundamentals; (3) partnerships with community colleges and four-year higher education institutions; and (4) expanding broadband access to reduce the digital divide.

Data from the Census Bureau and the Bureau of Labor Statistics suggest that the U.S. economy potentially missed out on $4.4 trillion dollars across all sectors as a result of lack of investment in women and minority businesses (Morgan Stanley, 2018). While this figure does not disaggregate data for the tech sector or for women of color, it suggests that there is a substantial financial loss to the sector related to underinvestment (McAlear et al., 2018). Potential strategies for reducing barriers to entry in the tech sector include increasing the number of women investors and investors of color, setting targets with measurable outcomes for investments in women of color, working to eliminate potential biases in screening of potential investment opportunities, and increasing targeted philanthropic investment in women of color.

CONCLUSIONS

Despite efforts to diversify the tech sector, women of color are disproportionately excluded at all levels of the tech workforce. Increasing the number of women of color in tech will require recruiting more individuals from more diverse sources as well as a culture shift within tech companies that welcomes the perspectives of women of color and recognizes the value they bring to the workforce. Improving the workplace experiences of women of color is equally important as increasing their number. For many women of color—even those who persist and advance in their careers—equity and inclusion issues in the workplace such as discrimination, isolation, and salary disparities continue to present a major ongoing challenge. More robust data collection, including intersectional data and data informed by the experiences of women of color in the workplace, is necessary to develop evidence-based strategies needed to increase diversity, equity, inclusion, and belonging.

Women of color are not a monolithic group. Intersectional strategies that explicitly and concretely address the challenges faced by women of color and other groups who encounter multiple, cumulative forms of bias and discrimination are a key component to improving recruitment, retention, and advancement. Many corporate diversity efforts focus on either race or gender rather than taking an intersectional approach, which often leads to the specific needs of women of color remaining unmet. In contrast, when companies set goals and track outcomes by both gender and race, they gain data-driven insights into the barriers women of color face, which allows them to target specific interventions to improve recruitment, retention, and advancement; reduce workplace bias; improve organizational culture; and make targeted investments in women of color. Moreover, these approaches appear to be beneficial to all employees and to the tech sector as a whole.

At the organizational level, leadership buy-in is key to the success of efforts to increase diversity, equity, and inclusion. Leaders with expertise in diversity, equity, and inclusion are particularly valuable assets to organizations trying to implement evidence-based strategies. Diversity, equity, and inclusion initiatives need adequate financial and staff resources as well as adequate time to be most effective at shifting industry/organizational culture and ensuring that diversity, equity, and inclusion priorities are connected to the business's overarching priorities and goals.

Different types of supports are needed at different stages of professional advancement in order to develop policies and practices that address gaps, improve retention and advancement, and increase equity. These supports might include advisory and mentoring programs as women of color transition into the workforce from higher education and ongoing support as they progress in their careers. The development of these practices and solutions should be informed by the experiences of women of color in order to best target their needs.

Data on women of color will be critical to fully understand how to leverage the skills of women of color. Innovation will depend on cultivating flexibility and creativity, investing in talent, and developing novel approaches to address challenges that currently impede the representation and success of women of color in tech.

RECOMMENDATIONS

The committee offers the following recommendations for increasing the recruitment, retention, and advancement of women of color in the tech industry.

RECOMMENDATION 4-1. To enhance the accuracy of data reporting, tech companies should disaggregate employment data by tech and non-tech positions, job titles, gender, and race/ethnicity—with particular attention to the intersection of race/ethnicity and gender—and make those data publicly available. Reports should include information about trends in recruitment, retention, and advancement of women of color.

Although some companies have been hesitant to disclose EEO-1 data to the public, many other companies recently began releasing company-wide demographic data. However, the majority of these reports classify women as a single underrepresented group despite vastly different trends among women from different racial and ethnic backgrounds. There remains a need for further transparency in order to fully understand the employment landscape of women of color in the tech industry. Demographic data play a critical role in measuring progress along with identifying areas where additional resources are needed to improve recruitment, retention, and advancement; benchmarking; and creating strategic plans for improving diversity, equity, and inclusion.

Companies, organizations, and researchers also need data on recruitment demographics, promotion rates, and attrition (both involuntary and voluntary) in order to identify inequities and remove structural and systemic barriers that contribute to women of color leaving high-tech positions. Transparency in reporting will promote accountability. Without these types of changes, it is unlikely that the tech sector will be able to reduce racial bias and discrimination.

RECOMMENDATION 4-2. Companies and organizations working within the tech sector should create pathways for women of color into leadership positions and create positions for diversity, equity, and inclusion professionals that are part of executive leadership.

Creating a diverse and inclusive organizational culture starts with leaders both in individual companies and across the industry who recognize their essential role in shaping organizational culture and DEI priorities. Diversity is a business imperative that requires the oversight and attention of senior executive management. Diversity, equity, and inclusion professionals within a company should have sufficient financial and human resources to support organizational goals in order to successfully implement research-based best practices with well-defined goals. They should also have direct access to other members of the leadership team, an opportunity to report on the status of progress, and they should be able to demonstrate measurable success in their role.

Increasing the number of women of color in leadership positions will improve equity in tech by building industry leadership that reflects the identities of the customers and communities the industry serves. It is important to note that continuity of leadership, sustained implementation of best practices, and consistency of metrics for assessing the success of DEI efforts are factors that can reduce the negative effects of frequent organizational change (e.g., lack of promotion or the need to rebuild credibility with teams, customers, and partners) and improve outcomes for women of color as they progress in their careers.

Women of color bring a wealth of skills, abilities, networks, and other cultural capital to the workplace. Evidence shows that companies with more diversity in leadership outperform companies with less diverse leaders. In addition, cultivating leaders who are women of color can improve innovation, increase recruitment of other women of color, and, in the long term, improve the diversity of the tech industry's talent pool.

RECOMMENDATION 4-3. Tech companies, with the assistance of a neutral central organization, should initiate an ongoing cross-sector coalition with each other as well as other stakeholders such as academic institutions—especially minority-serving institutions (e.g., historically Black colleges or universities, Hispanic-serving institutions, and tribal colleges and universities)—and professional societies. This collective would allow member organizations and

institutions to connect with each other with the goal of supporting current and future women of color in tech and promoting effective recruitment, retention, and advancement strategies for women of color in tech across all entities.

To create new solutions to improve diversity, equity, and inclusion, a collective approach to problem solving that facilitates the development of partnerships across the tech ecosystem could be a successful way to address the underrepresentation of women of color in tech. As other sectors increasingly utilize and develop new technologies, the tech sector will continue to expand and evolve. Although some collectives of tech companies already exist, a cross-sector, collective approach to strategic planning implemented in collaboration with a neutral, well-resourced central organization will help industry, higher education institutions, and other stakeholders (e.g., organizations working to create alternative pathways into tech and policy making) to increase accountability, share data, and leverage their strengths to develop strategies for improving policies and practices that improve outcomes for women of color as they transition from higher education into the workforce and as they advance in their careers in tech.

RECOMMENDATION 4-4. Tech companies should expand employment options that promote work-life balance such as remote work, flexible work hours, parental and other family leave, and career counseling as a strategy to improve retention and advancement and expand recruitment of women of color.

There are increased opportunities for recruiting and retaining a diverse workforce when companies implement practices that facilitate balance between work and home life. Although many companies within the tech sector have implemented flexible work policies, employees' opportunities to advance may sometimes be limited when they take full advantage of such policies. Although flexible work policies that promote work-life balancing have been shown to benefit both men and women, evidence shows that women—and women of color, in particular—are more negatively affected by the absence of these types of policies. Women shoulder a disproportionate burden of household management, childcare, and other caregiving. Research shows that implementing flexible work policies for all employees fosters more equitable participation in the workforce and increases opportunities for retention and advancement. Flexible work policies such as remote work may also be a valuable recruitment tool for attracting new employees who are women of color and allowing them the option to remain in geographic regions where they have easier access to family, community, and other support networks and resources.

REFERENCES

Allyn, B. 2020. *Ousted Black Google Researcher: 'They Wanted To Have My Presence, But Not Me Exactly.'* https://www.npr.org/2020/12/17/947719354/ousted-black-google-researcher-they-wanted-to-have-my-presence-but-not-me-exactl.

Alter, C. 2015. Intel pledges $300 million to increase workforce diversity. *Time.* January 7. https://time.com/3657636/intel-300-million-diversity-pledge/.

AnitaB.org. 2017. Advancing women technologists into positions of leadership: Strategies for cultivating confident women leaders. https://anitab.org/wp-content/uploads/2020/08/advancing-women-technologists-leaders.pdf.

AnitaB.org. 2018. Key programs help IBM retain and promote women tech leaders. https://anitab.org/profile/two-key-programs-help-ibm-retain-promote-women-tech-leaders/.

AnitaB.org. 2019a. Closing the diversity gap. https://anitab.org/wp-content/uploads/2020/08/AnitaB-TEF2019-CaseStudy-CapitalOne-FINALv02.pdf.

AnitaB.Org. 2019b. *Top companies for women technologists: Key findings and insights.* https://anitab.org/wp-content/uploads/2020/08/2019-Top-Companies-Insights-Report.pdf.

Ashcraft, C., B. McLain, and E. Eger. 2016. *Women in tech: The facts.* Boulder, CO: National Center for Women and Information Technology. https://ncwit.org/resource/thefacts/.

Bond, S. 2021. Senate Democrats Urge Google to Investigate Racial Bias in Its Tools and the Company. https://www.npr.org/2021/06/02/1002525048/senate-democrats-to-google-investigate-racial-bias-in-your-tools-and-company.

Bort, J. 2020. 2 Black women publicly resigned from Pinterest, saying they faced humiliation and retaliation and were passed over for promotion. *Business Insider,* June 15. https://www.businessinsider.com/two-black-women-publicly-explain-why-they-resigned-from-pinterest-2020-6.

Brown, D. 2017. Reflections on Intel's diversity and inclusion journey: 2016 diversity and inclusion annual report. *Intel Blog,* February 28. https://blogs.intel.com/csr/2017/02/annual-report-2016/#gs.z3y5ft.

Brown, S. 2020. The COVID-19 crisis continues to have uneven economic impact by race and ethnicity. *Urban Wire,* July 1. https://www.urban.org/urban-wire/covid-19-crisis-continues-have-uneven-economic-impact-race-and-ethnicity.

Carson, E. 2020. Intel offers another reminder working women had it tough in 2020. CNET, December 15. https://www.cnet.com/news/intel-offers-another-reminder-working-women-had-it-tough-in-2020/.

Catalyze Tech Working Group. 2021. The Act Report: Action to catalyze tech, a paradigm shift for DEI. Washington, DC: The Aspen Institute and Snap Inc. https://www.actreport.com/wp-content/uploads/2021/10/The-ACT-Report.pdf.

Connor, M. 2017. Tech still doesn't get diversity. Here's how to fix it. *Wired,* February 8. https://www.wired.com/2017/02/tech-still-doesnt-get-diversity-heres-fix/.

Crenshaw, K. 1991. Mapping the margins: Intersectionality, identity politics, and violence against women of color. *Stanford Law Review* 43(6):1241-1299.

DBP (Diversity Best Practices). 2009. Budget and staffing. In *Diversity Primer.* https://www.diversitybestpractices.com/sites/diversitybestpractices.com/files/import/embedded/anchors/files/diversity_primer_chapter_07.pdf.

DBP. March 16, 2018. Employee resource groups. https://www.diversitybestpractices.com/employee-resource-groups.

DBP. March 24, 2020. Diversity dialogue with Rainia L. Washington, vice president, global diversity and inclusion at Lockheed Martin. https://www.diversitybestpractices.com/diversity-dialogue-with-rainia-l-washington-vice-president-global-diversity-and-inclusion-at.

Delgado-Olsen, A. 2020. Addressing underrepresentation of women of color in tech: Perspectives of Native American women in computing. Presentation to the Committee on Addressing the Underrepresentation of Women of Color in Tech, June 4, 2020, Washington, DC.

Deloitte LLP. 2017. *Unleashing the power of inclusion: Attracting and engaging the evolving workforce.* https://www2.deloitte.com/content/dam/Deloitte/us/Documents/about-deloitte/us-about-deloitte-unleashing-power-of-inclusion.pdf.

Deloitte LLP. 2019a. *The bias barrier: Allyships, inclusion, and everyday behaviors.* https://www2.deloitte.com/content/dam/Deloitte/us/Documents/about-deloitte/us-inclusion-survey-research-the-bias-barrier.pdf.

Deloitte LLP. 2019b. *Uncovering talent: A new model of inclusion.* https://www2.deloitte.com/content/dam/Deloitte/us/Documents/about-deloitte/us-about-deloitte-uncovering-talent-a-new-model-of-inclusion.pdf.

Dickenson, S. 2021. What is allyship? *Communities,* January 28. National Institutes of Health. https://www.edi.nih.gov/blog/communities/what-allyship.

DuBow, W., and J. J. Gonzalez. 2020. NCWIT scorecard: The status of women in technology. Boulder, CO: National Center for Women and Information Technology.

Duffy, K. 2020. A Black ex-Googler claimed she was told by a manager that her Baltimore-accented speech was a 'disability' and later fired. *Business Insider,* December 22. https://www.businessinsider.com/google-fired-employee-diversity-recruiter-baltimore-accent-was-disability-2020-12.

Dwoskin, E. 2020. Complaint alleges that Facebook is biased against black workers. *The Washington Post,* July.

EEOC (Equal Employment Opportunity Commission). 2015. *Diversity in high tech.* https://www.eeoc.gov/special-report/diversity-high-tech.

Estrada, S. 2020. Intel plans to double number of women, underrepresented groups in leadership. *HR Dive,* May 18. https://www.hrdive.com/news/intel-plans-to-double-number-of-women-underrepresented-groups-in-leadershi/578122/.

Gee, B., and D. Peck. 2017. *The illusion of Asian success: Scant progress for minorities in cracking the glass ceiling from 2007–2015.* New York: Ascend. https://static1.squarespace.com/static/5e8bce29f730fc7358d4bc35/t/5fdd1f1ae4a5c90ed63228c4/1608326939357/The-Illusion-of-Asian-Success.pdf.

Griffin, Burge, Goldman, et al. 2021. How an Industry-Academic Partnership Addressed Gaps in Undergraduate CS Education (under review).

Gupta, A. H., and R. Tyshyan. 2021. 'You're the problem': When they spoke up about misconduct, they were offered mental health services. *The New York Times,* July 28. https://www.nytimes.com/2021/07/28/us/google-workplace-complaints-counseling.html.

Hewlett, S. A., C. B. Luca, L. J. Servon, L. Serbin, P. Shiller, E. Sosnovich, and K. Sumberg. 2008. *The Athena factor: Reversing the brain drain in science, engineering, and technology.* Harvard Business Review research report. Cambridge, MA. https://www.researchgate.net/publication/268325574_By_RESEARCH_REPORT_The_Athena_Factor_Reversing_the_Brain_Drain_in_Science_Engineering_and_Technology.

Hewlett, S. A., M. Marshall, and L. Sherbin. 2013. How diversity can drive innovation. *Harvard Business Review* 91(12):30.

Hewlett, S. A., L. Sherbin, F. Dieudonne, C. Fargnoli, and C. Fredman. 2014. Athena factor 2.0: Accelerating female talent in science, engineering, & technology. Center for Talent Innovation. https://coqual.org/reports/athena-factor-2-0-accelerating-female-talent-in-science-engineering-technology/.

Hodari, A. K., Ong, L. T. Ko, and J. M. Smith. 2016. Enacting agency: The strategies of women of color in computing. *Computing in Science and Engineering* 18(3):58-68.

HPC Wire. 2019. Intel releases 2019 diversity and inclusion report. https://www.hpcwire.com/off-the-wire/intel-releases-2019-diversity-and-inclusion-report/.

Huaranca Mendoza. 2020. Presentation to the Committee on Addressing the Underrepresentation of Women of Color in Tech, May 15, 2020, Washington, DC.

Ibarra, H., N. Carter, and C. Silva. 2010. Why men still get more promotions than women. *Harvard Business Review* 88(9):80-85.

IBM (International Business Machines Corporation). 2017. Empowering women's success in technology: IBM's commitment to inclusion. https://www.ibm.com/employment/inclusion/downloads/empowering_women_in_tech_ibm_case_study.pdf.

IBM. 2020. *IBM 2020 Diversity and inclusion report.* https://www.ibm.com/impact/be-equal/pdf/IBM_Diversity_Inclusion_Report_2020.pdf.

Intel. 2019. Intel 2017, 2018 and 2019 EEO-1 pay data. https://www.intel.com/content/www/us/en/diversity/2017-2019-eeo-1-pay-disclosure-report.html.

Intel. 2020. *Corporate responsibility at Intel. 2019-2020 report.* http://csrreportbuilder.intel.com/pdfbuilder/pdfs/CSR-2019-20-Full-Report.pdf.

Kang, C., and T. Frankel. 2015. Silicon Valley struggles to hack its diversity problem. *The Washington Post*, July 16. https://www.washingtonpost.com/business/economy/silicon-valley-struggles-to-hack-its-diversity-problem/2015/07/16/0b0144be-2053-11e5-84d5-eb37ee8eaa61_story.html.

Kapin, A. 2020. 51 VCs who want to invest in women, Black and Latinx, and LGBTQ+ founders. *Forbes*, December 12. https://www.forbes.com/sites/allysonkapin/2020/12/10/51-vcs-who-want-to-invest-in-women-black-and-latinx-and-lgbtq-founders.

Kapor Center. 2018. Statistics on women of color in computing provide the tech industry with key areas to target for transformation. September 12. https://women2.com/2018/09/12/women-in-computing-stats/.

Kerber, R., and S. Jessop. 2020. American companies facing pressure to reveal data on diversity of employees. *Insurance Journal*, July 6. https://www.insurancejournal.com/news/national/2020/07/06/574380.htm.

Kvasny, L., E. M. Trauth, and A. J. Morgan. 2009. Power relations in IT education and work: The intersectionality of gender, race, and class. *Journal of Information, Communication and Ethics in Society* 7(2-3):96-118.

Lampkin, S. 2020. Addressing underrepresentation of women of color in tech. Presentation to the Committee on Addressing the Underrepresentation of Women of Color in Tech, May 16, 2020, Washington, DC.

Lean In and McKinsey. 2020. *Women in the workplace, 2020.* https://wiw-report.s3.amazonaws.com/Women_in_the_Workplace_2020.pdf.

Lee, B. Y. 2020. Fostering diversity and inclusion at Uber. Presentation to the Committee on Addressing the Underrepresentation of Women of Color in Tech, May 15, 2020, Washington, DC.

Liu, S., and J. Parilla, 2019. Is America's Seed Fund investing in women- and minority-owned businesses? https://www.brookings.edu/blog/the-avenue/2019/06/04/is-americas-seed-fund-investing-in-women-and-minority-owned-businesses.

Lockheed Martin. 2016. *2016 Global diversity and inclusion annual report.* Bethesda, MD. https://www.lockheemartin.com/content/dam/lockheed-martin/eo/documents/globaldiversityinclusion/2016-global-diversity-inclusion-report.pdf.

Lockheed Martin. 2019. *ALL-INclusive transforming for impact: 2019 Global diversity and inclusion annual report.* Bethesda, MD. https://www.lockheedmartin.com/content/dam/lockheed-martin/eo/documents/globaldiversityinclusion/global-diversity-inclusion-annual-report-2019.pdf.

Malcom, S. M., P. Q. Hall, and J. W. Brown. 1976. *The double bind: The price of being a minority woman in science.* Publication 76-R-3. Washington, DC: American Association for the Advancement of Science. http://web.mit.edu/cortiz/www/Diversity/1975-DoubleBind.pdf.

Mallick, M. 2020. Do you know why your company needs a chief diversity officer? *Harvard Business Review*, September 11. Cambridge, MA. https://hbr.org/2020/09/do-you-know-why-your-company-needs-a-chief-diversity-officer.

Martinez-Cola, M. 2020. Collectors, nightlights, and allies, oh my! White mentors in the academy. Presentation to the Committee on Addressing the Underrepresentation of Women of Color in Tech, June 4, 2020, Washington, DC.

McAlear, F., A. Scott, K. Scott, and S. Weiss. 2018. Women of color in computing. Data brief. https://www.wocincomputing.org/wp-content/uploads/2018/08/WOCinComputingDataBrief.pdf.

McClain, B., C. Ashcraft, and J. Esch. 2021. Powertilt: Examining power, influence, and the myth of meritocracy within technology teams. Boulder, CO: NCWIT. https://www.ncwit.org/sites/default/files/resources/powertilt-final022621.pdf.

McGregor, J. 2020. Urged to back up pledges for racial justice, 34 major firms commit to disclose government workforce data. *The Washington Post*, September 29. https://www.washingtonpost.com/business/2020/09/29/corporate-diversity-data-pledge/.

McKinsey & Company. 2015. *Why diversity matters*. https://www.mckinsey.com/business-functions/organization/our-insights/why-diversity-matters.

McKinsey & Company. 2018. *Delivering through diversity*. https://www.mckinsey.com/~/media/mckinsey/business%20functions/organization/our%20insights/delivering%20through%20diversity/delivering-through-diversity_full-report.ashx.

McKinsey & Company. 2020. *Diversity wins: How inclusion matters*. May 19. https://www.mckinsey.com/featured-insights/diversity-and-inclusion/diversity-wins-how-inclusion-matters.

McNeely, C. L., and K. H. Fealing. 2018. Moving the needle, raising consciousness: The science and practice of broadening participation. *American Behavioral Scientist,* 62(5):551-562. https://doi.org/10.1177/0002764218768874.

Metcalf, H., D. Russell, and C. Hill. 2018. Broadening the science of broadening participation in STEM through critical mixed methodologies and intersectionality frameworks. *American Behavioral Scientist* 62(5):580-599.

Montañez, R. 2021. LinkedIn is now paying ERG leaders; this is huge in the battle against burnout. *Forbes,* June 3. https://www.forbes.com/sites/rachelmontanez/2021/06/03/linkedin-is-now-paying-erg-leaders-this-is-huge-in-the-battle-against-burnout.

Morris, S. E. 2021. LinkedIn joins the bandwagon by compensating ERG leaders for culture Impacts. *Forbes,* June 17. https://www.forbes.com/sites/simonemorris/2021/06/17/linkedin-joins-the-bandwagon-by-compensating-erg-leaders-for-culture-impacts.

Morgan Stanley. 2018. *The growing market investors are missing: The trillion-dollar case for investing in female and multicultural entrepreneurs*. https://www.morganstanley.com/pub/content/dam/msdotcom/mcil/growing-market-investors-are-missing.pdf.

Murphy, M. C., K. M. Kroeper, and E. M. Ozier. 2018. Prejudiced places: How contexts shape inequality and how policy can change them. *Policy Insights from the Behavioral and Brain Sciences* 5(1):66-74.

NASEM (National Academies of Sciences, Engineering, and Medicine). 2019. *The science of effective mentorship in STEMM*. Washington, DC: The National Academies Press. https://doi.org/10.17226/25568.

NASEM. 2021. *The impact of COVID-19 on the careers of women in academic sciences, engineering, and medicine*. Washington, DC: The National Academies Press.

NCWIT (National Center for Women and Information Technology). 2020. Highlights from the NCWIT scorecard: Indicator data showing the participation of girls and women in computing. Boulder, CO. https://www.ncwit.org/sites/default/files/resources/ncwit_scorecard_data_highlights_10082020.pdf.

Nordli, B. 2019. For women of color in tech, it's "hard to grow" without representation. July 29. https://builtin.com/women-tech/women-color-tech-inclusion.

Ong, M., C. Wright, L. Espinosa, and G. Orfield. 2011. Inside the double bind: A synthesis of empirical research on undergraduate and graduate women of color in science, technology, engineering, and mathematics. *Harvard Educational Review* 81(2):172-209.

Ong, M., J. M. Smith, and L. T. Ko. 2018. Counterspaces for women of color in STEM higher education: Marginal and central spaces for persistence and success. *TEA Journal of Research Science Teaching* 55(2):206-245.

Pace, C. 2018. How women of color get to senior management. *Harvard Business Review*, August 31. Cambridge, MA. https://hbr.org/2018/08/how-women-of-color-get-to-senior-management.

Parker, M. 2020. Google diversity. Presentation to the Committee on Addressing the Underrepresentation of Women of Color in Tech, May 15, 2020, Washington, DC.

Peck, D. 2020. Ascend leadership. Presentation to the Committee on Addressing the Underrepresentation of Women of Color in Tech, May 14, 2020, Washington, DC.

Pawley, A. 2020. Addressing the small-N problem. Presentation to the Committee on Addressing the Underrepresentation of Women of Color in Tech, June 3, 2020, Washington, DC.

Pivotal Ventures and McKinsey & Company. 2018. *Rebooting representation.* https://127j5241 bcgw285yu54bgh7m-wpengine.netdna-ssl.com/wp-content/uploads/Rebooting-Representation-Report.pdf.

Reuters. 2018. *Uber agrees to settle California discrimination lawsuit for $10 million.* https://www.reuters.com/article/us-uber-lawsuit/uber-agrees-to-settle-california-discrimination-lawsuit-for-10-million-idUSKBN1H40A7.

Scott, A., F. Klein, F. McAleaer, A. Martin, and S. Koshy. 2018. *The leaky tech pipeline: A comprehensive framework.* Oakland, CA: Kapor Center for Social Impact. http://www.leakytechpipeline.com/wp-content/themes/kapor/pdf/KC18001_report_v6.pdf.

Shah, S. 2020. *Breaking through, rising up: Strategies for propelling women of color in technology.* Brooklyn, NY: NPower. https://www.npower.org/wp-content/uploads/2020/06/NPower_40x22_report_060120_Single-Pages.pdf.

Sherbin, L., and R. Rashid. 2017. Diversity doesn't stick without inclusion. *Harvard Business Review.* February 1. https://hbr.org/2017/02/diversity-doesnt-stick-without-inclusion.

Shook, E., and J. Sweet. 2019. *Getting to equal 2019: Creating a culture that drives innovation.* Accenture. https://www.accenture.com/_acnmedia/Thought-Leadership-Assets/PDF/Accenture-Equality-Equals-Innovation-Gender-Equality-Research-Report-IWD-2019.pdf.

Simonite, T. 2021. What really happened when Google ousted Timnit Gebru. *Wired*, June 8. https://www.wired.com/story/google-timnit-gebru-ai-what-really-happened.

Skervin, A. 2015. Success factors for women of color information technology leaders in corporate America. Dissertation. Minneapolis, MN: Walden University.

Solórzano, D. M. C., and Y. Tara. 2003. Critical race theory, racial microaggressions, and campus racial climate: The experiences of African American college students. *Educational Administration Abstracts* 38(1):3-139.

Tiku, N. 2020. Tech companies are asking their Black employee groups to fix Silicon Valley's race problem—often for free. *Washington Post*, June 26. https://www.washingtonpost.com/technology/2020/06/26/black-ergs-tech/.

Tiku, N. 2021. Google's approach to historically Black schools helps explain why there are few Black engineers in Big Tech. *Washington Post*, March 4. https://www.washingtonpost.com/technology/2021/03/04/google-hbcu-recruiting.

van Nierop, R. 2020. Latinas in Tech. Presentation to the Committee on Addressing the Underrepresentation of Women of Color in Tech. May 14, 2020, Washington DC.

Varma, R. 2018. U.S. science and engineering workforce: Underrepresentation of women and minorities. *American Behavioral Scientist* 62(5):692-697.

Vasel, K. 2019. Intel was losing employees. So it created an anonymous hotline to help unhappy workers. *CNN*, May 23. https://www.cnn.com/2019/05/23/success/intel-warmline-employee-retention.

Washington, G. 2020. Retention of women of color students in tech. Presentation to the Committee on Addressing the Underrepresentation of Women of Color in Tech. April 7, 2020, Washington DC.

Williams, J. C. 2015. The 5 biases pushing women out of STEM. *Harvard Business Review*, March 24. https://hbr.org/2015/03/the-5-biases-pushing-women-out-of-stem.

Williams, J. C., K. W. Phillips, and E. V. Hall. 2016. Tools for change: Boosting the retention of women in the STEM pipeline. *Journal of Research in Gender Studies* 6(1):11-75.

Women Who Tech. 2020. *The state of women in tech and startups*. https://womenwhotech.com/data-and-resources/state-women-tech-and-startups.

5

The Role of Government in Addressing the Underrepresentation of Women of Color in Tech

As described in the earlier chapters of this report, the education and careers of women of color in technology and computing fields are impeded by both structural racism and sexism, which affect women of color in complex, cumulative, and intersecting ways as the discrimination they experience is compounded by their multiple social identities (Crenshaw, 1989). Government institutions have created and perpetuated structural and institutional racism and sexism in myriad ways (e.g., slavery, segregation, internment, disenfranchisement based on gender and race, and discriminatory housing, banking, and educational practices). While the government has acted to try to mitigate the impact of structural racism and sexism in American society, in general and in science, technology, engineering, and mathematics (STEM) fields in particular, in many cases such efforts have failed to take an intersectional approach.[1] Without such an approach, women of color often fail to benefit from government efforts to the same degree as white women or men from underrepresented groups.

Take the historic example provided by the fight for voting rights in the United States, both the white women who led the women's suffrage movement and the men of color who fought against disenfranchisement on the basis of race often excluded women of color from these civil rights efforts (Bailey, 2020). The passage of the 15th Amendment to the Constitution in 1870, which prevented states from prohibiting the right to vote on the basis of "race, color, or previous

---

[1] While the focus of this report is on tech fields, many of the government-level efforts and interventions that could benefit women of color in tech would also be beneficial for women of color pursuing STEM fields more generally. For this reason, this chapter often highlights efforts related to STEM in addition to those focused specifically on tech.

condition of servitude," enfranchised men of color, at least on paper,[2] but did not grant women the right to vote. The passage of the 19th Amendment in 1920, which granted women the right to vote and, in doing so, expanded professional and educational opportunities for women including in STEM fields, also failed to fully enfranchise women of color. Despite the contributions of women of color over decades to fight for women's suffrage, the same discriminatory obstacles and policies that prevented men of color from voting after the passage of the 15th Amendment also effectively withheld the right to vote from women of color. It was not until the passage of the Voting Rights Act in 1965, more than 40 years after the passage of the 19th Amendment, that women of color were able to more fully exercise their right to vote (Bleiweis et al., 2020).

In her 2020 opinion editorial in *Science*, historian of science and the co-chair of the committee authoring this report, Evelynn Hammonds, explained how this historical backdrop established a "legacy of exclusion" that persists today, both in society at large and in STEM fields. Hammonds wrote:

> Many of these (white) women joined in the suffrage movement, with the idea that the vote would help to advance their progress in scientific fields, but they often failed to confront their own exclusionary practices, particularly those surrounding race. In not advocating for voting rights for all women, they helped to support the segregation of scientists of color within scientific institutions, especially female scientists of color. Indeed, little was done by leading scientists to address issues of race or the representation of women of color in science until after World War II. Even after decades of efforts to increase the diversity of the U.S. scientific workforce, we are still struggling with this legacy of exclusion today (Hammonds, 2020).

Historical examples that illustrate this legacy of exclusion underscore the point made throughout this report that unless policies, practices, programs, and individuals embrace an intersectional approach in efforts to promote diversity, equity, and inclusion in institutions of higher education, in government, and in the tech workforce, women of color will likely continue to fail to fully benefit from these efforts.

In this chapter, the committee reviews efforts by government to address institutional obstacles facing women of color in tech; calls upon government to be more intentional about taking an intersectional approach to efforts aimed at supporting greater equity, diversity, and inclusion in tech (Box 5-1); and calls for additional action by government institutions to promote transparency and accountability among tech companies, institutions of higher education, and in the government itself, including in government funded laboratories. The chapter

---

[2] Even after the passage of the 15th Amendment, many men of color were often prevented from voting by a range of discriminatory policies and practices at the state level, such as poll taxes and literacy tests.

BOX 5-1
What Does It Mean to Take an Intersectional Approach?

An intentionally intersectional approach can take many forms. In essence, it means that government efforts focused on promoting diversity, equity, and inclusion in technology and computing fields must consider the unique experiences of women of multiple marginalized identities and be explicit about doing so. Taking an intersectional approach in government efforts to promote diversity, equity, and inclusion may mean disaggregating quantitative data by race and gender or instructing grant recipients to carry out qualitative data collection in order to better understand the experiences of women of color in tech. Government efforts that use language such as "women and minorities" fail to account for intersectionality because the use of this language treats gender and race as distinct rather than intersecting.

---

highlights opportunities for organizations working to promote greater equity, diversity, and inclusion in tech education and careers to form high-impact partnerships to expand their sphere of influence.

EFFORTS BY GOVERNMENT TO MITIGATE INSTITUTIONAL OBSTACLES FOR WOMEN OF COLOR

In the section below the committee reviews efforts by Congress and federal agencies to support women of color pursuing tech education and careers and attempts by these government institutions to remove systemic barriers to access and equity in these fields. The committee identifies progress that has been made by these efforts, including a recent, growing appreciation for the importance of taking an intersectional approach to such efforts, as well as shortcomings and lessons learned from past efforts. This overview forms the basis of several recommendations targeted at government that appear at the end of the chapter.

Legislative Efforts

Notable historical examples of legislative efforts to address structural racism and sexism in education and the workplace include the passage and amendment of Title VII, originally part of the Civil Rights Act of 1964, which "prohibits employment discrimination based on race, color, religion, sex, and national origin" and the passage of Title IX as part of the Education Amendments of 1972, which states that "no person in the United States shall, on the basis of sex, be excluded from participation in, be denied the benefits of, or be subjected to discrimination under any education program or activity receiving federal financial assistance."

It is worth pointing out that these laws focus on gender and race separately and do not consider intersectionality.

In its research, the committee found many examples of past legislative efforts aimed at addressing the underrepresentation of women and minorities in STEM fields. In the vast majority of cases, such legislative efforts did not emphasize the concept of intersectionality or account for the ways that women of color experience heightened forms of bias, discrimination, and harassment (see Chapters 2 and 3). Rather, the language in legislation has tended to focus either on women *or* on people from underrepresented groups in STEM.

Nevertheless, it is encouraging to see that two recent bills introduced by the U.S. House Committee on Science, Space, and Technology, under the leadership of Chairwoman Eddie Bernice Johnson, include language that acknowledges intersectionality. One, the proposed STEM Opportunities Act, "provides for guidance, data collection, and grants for groups historically underrepresented in science, technology, engineering, and mathematics education at institutions of higher education and at federal agencies." Though the term "intersectionality" is not specifically used in the draft language for the STEM Opportunities Act, in section 4, "Collection and Reporting of Data on Federal Research Grant," the legislation calls for disaggregation of data "cross-tabulated by race, ethnicity, gender, and years since completion of doctoral degree." Furthermore, in section 7b, "Workshops to Address Cultural Barriers to Expanding the Academic and Federal STEM Workforce," the bill includes language tasking agencies with ensuring that the workshops provide "a discussion of the unique challenges faced by different underrepresented groups, including minority women, minority men, persons from rural and underserved areas, persons with disabilities, gender and sexual minority individuals, and first-generation graduates in research." A second example, the "Combating Sexual Harassment in Science Act of 2019," which seeks to address "sexual harassment and gender harassment in the science, technology, engineering, and mathematics fields by supporting research regarding such harassment and efforts to prevent and respond to such harassment" calls upon the National Science Foundation (NSF) to fund "research on the sexual harassment and gender harassment experiences of individuals in underrepresented or vulnerable groups, including racial and ethnic minority groups, disabled individuals, foreign nationals, sexual- and gender-minority individuals, and others." These examples of legislative language that acknowledge, explicitly or implicitly, that it is important to examine the intersection of race and gender in efforts to address underrepresentation in STEM are, in this committee's opinion, an important step in the right direction. In its research, the committee was also struck by how often legislation related to diversity, equity, and inclusion in STEM has been introduced by various congressional committees but how rarely it has been passed into law. For instance, in the case of the STEM Opportunities Act described above, similar legislation has been introduced in every Congress since 2007 (U.S. House of Representatives, 2021).

Federal Agencies' Efforts

Federal agencies such as NSF, the Department of Defense (DoD), the National Institute of Standards and Technology (NIST), and the National Aeronautics and Space Administration (NASA) have implemented programs intended to address the underrepresentation of specific groups in tech. In the section below, we offer a brief overview of programs and initiatives under way across a range of federal agencies. In general, however, the committee observed that information about federal agencies' direct or indirect support of women of color in tech is widely dispersed and inconsistently distributed on various agency websites, which made gaining a complete understanding and an accurate record of these investments challenging. The one exception was NSF, for which the annual budget request to Congress provides a compilation of the agency's efforts each year to support broadening participation (NSF, 2020b). This is an incredibly useful resource for those seeking to understand NSF's investments in equity, diversity, and inclusion in STEM fields. Such a resource is, unfortunately, not available for all federal agencies, which can make ascertaining the ways that other agencies are investing in such efforts difficult. Table 5-1 describes the list of programs NSF reported in the 2021 budget request to Congress that focus[3] on broadening participation and offers information on the extent to which individual programs take an intersectional approach or support women of color in tech, specifically. The committee sees an important need for additional federal agencies to create a similar record of account for the ways in which they are investing in efforts to support a diverse and inclusive STEM workforce (see Recommendation 5-2).

Among those efforts at NSF most directly relevant to tech education and careers are the range of programs and initiatives led by the Directorate of Computer and Information Science and Engineering (CISE). CISE has taken steps to support broadening participation in tech. For example, in 2017, CISE launched a pilot program on broadening participation in computing. The effort encourages the inclusion of broadening participation plans as part of proposal requirements for a subset of granting programs in the directorate. The directorate has "developed a review and feedback process to ensure that these…plans are meaningful, include concrete metrics for success, and that progress toward goals is included as part of project annual reports." In September 2021, CISE released updated guidelines for the program, which included requirements that proposals to the CISE core research programs requesting between $600,001 and $1,200,000 in funding include broadening participation plans at the time of submission. In its information gathering, the committee noted that the pilot program for broadening participation in computing did not require prospective grantees to include the broadening participation plan in the proposal at the time of review, but rather

---

[3] In addition to its programs that "focus" on broadening participation, NSF also reports to Congress an additional list of programs that "emphasize" broadening participation.

at the time of award. As the CISE directorate continues to shape this effort, the committee would encourage the directorate to examine and evaluate the impact of this process on the quality of the broadening participation plans and the degree to which grantees follow through with their stated plans, as indicated by the annual reports submitted to NSF. Additional CISE efforts that support broadening participation in tech education and careers include the Computer and Information Science and Engineering Minority-Serving Institutions Research Expansion Program,[4] the Broadening Participation in Computing Alliance, and Computer Science for All (NSF, 2020a, 2020b, 2021).[5]

In its research, the committee found that many of the solicitations (i.e., funding announcements) for programs at NSF explicitly emphasize intersectionality (see Table 5-1). This is noteworthy because, historically, this has not always been the case. For example, NSF's ADVANCE program (Organizational Change for Gender Equity in STEM Academic Professions), launched in 2001, which had as its goal "to increase the representation and advancement of women in academic science and engineering careers, thereby contributing to the development of a more diverse science and engineering workforce," did not initially take an intersectional approach. Research has shown that white women have benefited predominantly from ADVANCE (NSF, 2019). In recent years, NSF has acknowledged this issue with the ADVANCE program and now includes a focus on intersectionality as a requirement for prospective ADVANCE grantees (Rosser et al., 2019). The 2020-2021 program solicitation for the ADVANCE program reads, "All NSF ADVANCE proposals are expected to use intersectional approaches in the design of systemic change strategies in recognition that gender, race, and ethnicity do not exist in isolation from each other and from other categories of social identity." This is an important step. The committee encourages NSF to monitor over time the impact of this intersectional approach on women of color at institutions that are recipients of ADVANCE grants.

The committee would also encourage NSF to emphasize intersectionality in its "broader impacts" merit review criteria in its core research programs. Currently, the "broader impacts" merit review criteria language does not acknowledge intersectionality but rather describes "full participation of women, persons with disabilities, and underrepresented minorities in science, technology, engineering, and mathematics." The committee would encourage NSF to consider revising this language to explicitly account for intersectionality (see Recommendation 5-1) and to collect disaggregated data to monitor the extent to which the core research programs at NSF are, or are not, supporting women of color in tech.

In addition to funding efforts by grantees to improve equity, diversity, and in-

---

[4] National Science Foundation. 2020. Computer and information science and engineering minority-serving institutions research expansion program (CISE-MSI Program). https://beta.nsf.gov/funding/opportunities/computer-and-information-science-and-engineering-minority-serving.

[5] National Science Foundation. 2021. Broadening participation in computing (BPC). https://www.nsf.gov/cise/bpc/.

Table 5-1 Programs at the National Science Foundation that Focus on Broadening Participation in STEM from 2021 Budget Request to Congress

Program Name	Program Description and Reference to Intersectionality
ADVANCE (Organizational Change for Gender Equity in STEM Academic Professions)	The NSF ADVANCE program provides grants to promote systemic change in support of equity and inclusion in the academic profession and workplaces. An intersectional approach is required for all proposals and awards. The solicitation reads: "All NSF ADVANCE proposals are expected to use intersectional approaches in the design of systemic change strategies in recognition that gender, race, and ethnicity do not exist in isolation from each other and from other categories of social identity." (https://www.nsf.gov/pubs/2020/nsf20554/nsf20554.htm)
Alliances for Graduate Education and the Professoriate	The AGEP program aims to increase the number of faculty from historically underrepresented groups through alliances between institutions of higher education, which NSF expects to "work collaboratively and use intersectional approaches in the design, implementation, and evaluation of systemic change strategies." (https://www.nsf.gov/pubs/2021/nsf21576/nsf21576.htm)
Alliances for Graduate Education and the Professoriate's Graduate Research Supplements	These supplements are available to institutions that currently or previously had an Alliances for Graduate Education and the Professoriate award and are to be used to support an individual doctoral student or master's student planning to pursue a Ph.D. (https://www.nsf.gov/pubs/2020/nsf20083/nsf20083.jsp)
Broadening Participation in Biology Fellowships	This program supports postdoctoral research fellowships in biology and is designed to provide opportunities for early-career scientists who are ready to assume independence in their research efforts and obtain training beyond their graduate education, gain research experience in collaboration with established scientists, and broaden their scientific horizons. The program includes a competitive area on broadening participation of groups underrepresented in biology that seeks "to increase the diversity of scientists explicitly at the postdoctoral level in biology, and thereby contribute to the future vitality of the nation's scientific enterprise." (https://www.nsf.gov/pubs/2020/nsf20602/nsf20602.htm)
Broadening Participation in Engineering	This program supports research aimed at providing "scientific evidence that engineering educators, employers, and policy makers need to make informed decisions to design effective programs that broaden the participation of persons from historically underrepresented groups in the engineering workforce." The solicitation states that this program "is particularly interested in research that employs intersectional approaches in recognition that gender, race, and ethnicity do not exist in isolation from each other and from other categories of social identity." (https://www.nsf.gov/pubs/2022/nsf22514/nsf22514.htm)

continued

Table 5-1 Continued

Program Name	Program Description and Reference to Intersectionality
Career-Life Balance	This program is primarily to support the career-life balance needs of individual NSF-funded principle investigators. For example, it "permit[s] the extension of NSF awards for researchers who take a leave of absence for dependent care responsibilities, as well as the use of NSF award funds to replace project personnel during a leave of absence." (https://www.nsf.gov/career-life-balance/)
Centers of Research Excellence in Science and Technology	These awards support minority-serving institutions through various avenues such as building research capacity, providing fellowships to individual postdocs, establishing partnerships with other entities like research institutions and K-12 schools. (https://www.nsf.gov/pubs/2018/nsf18509/nsf18509.htm)
Excellence Awards in Science and Engineering	This program includes the Presidential Awards for Excellence in Science, Mathematics and Engineering Mentoring and the Presidential Awards for Excellence in Mathematics and Science Teaching, which "fall under NSF's core value of inclusiveness—seeking and embracing contributions from all sources, including underrepresented groups, regions, and institutions." The awards are available to both individuals and organizations and are administered by NSF on behalf of the White House Office of Science and Technology Policy. Each awardee receives a certificate signed by the President of the United States and a $10,000 award from NSF, and is honored at an award ceremony in Washington, DC. (https://www.nsf.gov/funding/pgm_summ.jsp?pims_id=5473)
Historically Black Colleges and Universities Undergraduate Program	This program "provides awards to strengthen STEM undergraduate education and research at historically Black colleges and universities." (https://www.nsf.gov/pubs/2020/nsf20559/nsf20559.htm)
Historically Black Colleges and Universities Excellence in Research	The HBCU-EiR program aims to strengthen research capacity at HBCUs by funding research projects aligned with NSF's research programs. The program was established in response to direction provided by the Senate Commerce and Justice, Science and Related Agencies Appropriations Subcommittee Report (Senate Report 115-139). (https://www.nsf.gov/pubs/2020/nsf20542/nsf20542.htm)

Table 5-1 Continued

Program Name	Program Description and Reference to Intersectionality
Improving Undergraduate STEM Education: Hispanic-Serving Institutions program	The goals of this program are to support the improved recruitment, retention, and graduation rate of undergraduate students at HSIs (at the baccalaureate and associates levels) and to improve the quality of undergraduate STEM education at HSIs, while taking into account the wide range of institutional contexts and characteristics of HSIs and the students they serve. The language from the solicitation reads: "proposers are encouraged to use an intersectional perspective in designing proposals across all tracks in the Hispanic-serving institutions program. An intersectional lens takes into consideration the interconnectedness of overlapping social identities, and can help shape a project's design and conceptualization of inclusivity to better serve students. More than 50% of Hispanic/Latino/a undergraduate students attend Hispanic-serving institutions, and an intersectional approach to meeting them where they are could significantly impact the diversity of undergraduate STEM degrees awarded and STEM professionals in the United States." (https://www.nsf.gov/pubs/2020/nsf20599/nsf20599.htm)
Inclusion across the Nation of Communities of Learners of Underrepresented Discoverers in Engineering and Science (NSF INCLUDES)	INCLUDES supports a wide range of projects, events, and networks support broadening participation in STEM. INCLUDES is anchored in "five design elements of collaborative infrastructure to achieve systemic change" through which partnering organizations work together. The five design elements are (1) shared vision, (2) partnerships, (3) goals and metrics, (4) leadership and communication, and (5) expansion, sustainability and scale. (https://www.nsf.gov/pubs/2020/nsf20569/nsf20569.htm)
Louis Stokes Alliances for Minority Participation	The LSAMP is an alliance-based program that focused on assisting higher education institutions in increasing the number of STEM undergraduate and graduate degrees awarded to historically underrepresented groups in STEM, particularly underrepresented minorities. The program provides funding for the higher education alliances to "implement comprehensive, evidence-based, innovative, and sustained strategies that ultimately result in the graduation of well-prepared, highly-qualified students from underrepresented minority groups who pursue graduate studies or careers in STEM." (https://www.nsf.gov/pubs/2020/nsf20590/nsf20590.htm)

continued

Table 5-1 Continued

Program Name	Program Description and Reference to Intersectionality
Partnerships for Research and Education in Materials	The Partnerships for Research and Education in Materials Research program "aims to enable, build, and grow partnerships between minority-serving institutions" and centers and/or facilities supported by the Division of Materials Research. The program seeks to increase recruitment, retention, and degree attainment by members of those groups most underrepresented in materials research, and at the same time support excellent research and education endeavors that strengthen such partnerships." (https://www.nsf.gov/pubs/2021/nsf21510/nsf21510.htm)
Directorate for Social, Behavioral, and Economic Sciences Postdoctoral Research Fellowships-Broadening Participation	This program supports postdoctoral training in the social, behavioral, and economic (SBE) sciences through two tracks—one that supports basic research and another specifically focused on broadening participation in these fields. The broadening participation track seeks to "prepare underrepresented SBE scientists and others who share NSF's diversity goals for positions of scientific leadership in academia, industry, and government." The solicitation specifically cites data from the National Center for Science and Engineering Statistics that show that "American Indians or Alaska Natives, Blacks or African-Americans, Hispanics, and Native Hawaiians or Pacific Islanders, in addition to individuals with disabilities, are underrepresented in the SBE sciences in the U.S." (https://www.nsf.gov/pubs/2018/nsf18584/nsf18584.htm)
Science of Broadening Participation	This program "uses the theories, methods, and analytic techniques of the social, behavioral, economic, and learning sciences to better understand the factors that enhance as well as the barriers that hinder our ability to expand participation in education, the workforce and major social institutions in society, including science, technology, engineering, and mathematics (STEM) and other sectors." (https://beta.nsf.gov/funding/opportunities/sbe-science-broadening-participation-sbe-sbp)
Tribal Colleges and Universities Program	This program "provides awards to tribal colleges and universities, Alaska Native–serving institutions, and Native Hawaiian–serving institutions to promote high quality science (including sociology, psychology, anthropology, economics, statistics, and other social and behavioral sciences as well as natural sciences), technology, engineering and mathematics education, research, and outreach." (https://www.nsf.gov/pubs/2018/nsf18546/nsf18546.htm)

clusion in tech, NSF is also funding social science research to fill knowledge gaps on the experiences of women of color in tech. The Social and Behavioral Sciences Directorate leads the Science of Broadening Participation Program, which "uses the theories, methods, and analytic techniques of the social, behavioral, economic, and learning sciences to better understand the factors that enhance, as well as the barriers that hinder, our ability to expand participation in education, the workforce, and major social institutions in society, including science, technology, engineering, and mathematics and other sectors." Among the research questions the Science of Broadening Participation Program seeks to address are

- What are the underlying psychological and social issues affecting the different participation and graduation rates of people who vary by gender, race, ethnicity, disability, and other statuses in education, both within STEM but also in other fields?
- What social, behavioral, or economic processes and mechanisms contribute to positive outcomes within education, the workforce, and major social institutions in society? Do those processes and mechanisms differ by gender, race, ethnicity, disability, and other statuses?
- What factors help promote and maintain the interest of youth from underrepresented groups in education, including STEM fields?
- What are the impacts of a diverse workforce on scientific productivity and innovation and the national economy (NSF, 2018)?

This table describes the list of programs NSF reported in the 2021 budget request to Congress that focus on broadening participation and offers information on language in the program solicitations related to intersectionality. In addition to its programs that focus on broadening participation, NSF also reports to Congress an additional list of programs that emphasize broadening participation.

DoD-led efforts focused on STEM education, enrichment, and workforce development include those that are part of the "DoD STEM" effort. The DoD STEM website describes a vision of "a diverse and sustainable Science, Technology, Engineering, and Mathematics (STEM) talent pool ready to serve our Nation and evolve the Department of Defense's competitive edge" and offers 78 opportunities with different intended audiences (early-career professionals, educators, graduates, students, volunteers), types of experiences (competitions, educator development, fellowships, internships or apprenticeships, programs, scholarships), and educational levels (ranging from kindergarten to graduate level) (U.S Department of Defense STEM, n.d.). While diversity and inclusion appears to be a cross-cutting theme of these DoD STEM efforts, it is not clear from the website whether the DoD STEM efforts take an intersectional approach or emphasize women of color. That said, one DoD STEM program that reports conducting intentional outreach to women of color and other underrepresented groups in tech is the Science, Mathematics, and Research for Transformation (SMART) Scholarships-for-Service Program. SMART offers "scholarships for undergraduate, master's, and doctoral students currently pursuing a degree in one of its 21 STEM disciplines" and "scholars receive full tuition, annual stipends, health insurance, experienced mentors, internships, and guaranteed employment with the DoD after graduation." The leadership of the SMART Program is working to raise awareness of the SMART Program at minority-serving institutions. The program is currently holding webinars with the theme "fostering a community of diversity" focusing on inclusion, and broadening awareness of possible career pathways in federal service by highlighting the experiences of scholars that include women of color who have pursued careers in tech through the SMART Program.

DoD also seeks to support diversity, equity, and inclusion through Special Emphasis Programs which "are management programs established to ensure equal employment opportunity for minorities, women and individuals with disabilities in various categories and occupations and in all organizational components" (U.S. Department of Defense Education Activity, 2021). Among the Special Emphasis Programs offered are a Federal Women's Program, a Hispanic Emphasis Program, a Disability Emphasis Program, an African-American or Black Emphasis Program, an American Indian Program, and an Asian Pacific Islander Program. It is unclear whether these specific programs consider intersectionality (NWBC, 2020).

NIST is working across multiple levels in the organization to try to support women of color in tech, including through affinity groups, student and teaching opportunities, trainings and seminars, and through the efforts of the Steering Group for Equity in Career Advancement, which consists of staff from across NIST who serve as trusted advisors to senior leaders. NIST's self-reported activities focused on recruitment of women of color are summarized in Table 5-2.

NASA also offers programming in support of women of color in tech through its Minority University Research and Education Project (MUREP). MUREP offers competitive awards to minority serving institutions, including Historically Black Colleges and Universities, Hispanic–Serving Institutions, Asian American and Native American Pacific Islander–Serving Institutions, Alaska Native and Native Hawaiian-Serving Institutions, American Indian Tribal Colleges and Universities, Native American–Serving Nontribal Institutions and other minority-serving institutions. The goal of the program is to "assist faculty and students in research and provide authentic STEM engagement related to NASA missions… and provide NASA-specific knowledge and skills to learners who have historically been underrepresented and underserved in STEM."[6]

While special programs to support and advance women of color in tech, such as those at NSF, DoD, NIST, and NASA described above, are critically im-

---

[6] See https://www.nasa.gov/stem/murep/about/index.html.

Table 5-2 Self-Reported Efforts by the National Institute of Standards and Technology to Support the Retention, Recruitment, and Advancement of Women of Color in Tech

Initiative/Activity	Description
Conference for Undergraduate Underrepresented Minorities in Physics	NIST hosted the third conference January 8-10, 2021, and added a high school component.
Department of Commerce group for Women in STEM	This group was created in 2018 by an African American female electrical engineer working at NIST. Over time the group has grown to more than 330 members across the agency.
Engineering Laboratory Diversity, Inclusion, and Belonging Council	The council catalyzes the recruitment, hiring, development, and retention of high-performing staff to achieve its vision of a culture of diversity and inclusion. It facilitates transparent and equitable processes and policies so that all individuals develop professionally and reach their full potential.
Grace Hopper Celebration of Women in Computing[1]	NIST hosts a booth at this event that highlights its measurement science research, guest researcher program, PREP, SURF, and any current USAjobs postings.
Historically Black Colleges and Universities STEM Alliance Seminar	This seminar series, launched in September 2020, focuses on the research and career journeys of African American scientists from NIST and is geared toward historically Black college and university STEM undergraduates at three partnering colleges, Savannah State University, Texas Southern, and Prairie View A&M.
Information Technology Laboratory Diversity Committee	This committee is a grassroots assembly formed to assist the staff and management of NIST's Information Technology Laboratory in working toward the vision and the goals of the NIST policy on diversity. Key activities include awards, a winter social, and a book club.
LinkedIn Recruiter	The use of modern and proactive recruiting tools like LinkedIn Recruiter is a best practice in both public and private sectors. NIST is running a pilot of LinkedIn Recruiter from January 25 through September 30, 2021, with the goal of building pipelines of highly qualified and diverse candidates for current and future job openings. Recruiters are currently conducting proactive searches and identifying candidates to ask to apply to NIST vacancies.
Maryland Quantum Alliance (Mid-Atlantic Quantum Alliance, n.d.)	This education/workforce committee, of which NIST staff are a part, includes a focus on students from underrepresented groups.

---

[1] See https://ghc.anitab.org/.

continued

Table 5-2 Continued

Initiative/Activity	Description
Maryland STEM Festival: STEM of Many Colors	NIST has a booth at this festival to disseminate information about its academic funding activities, particularly SURF and SHIP, and to discuss STEM careers with students from underrepresented groups.
Milligan-May Symposium at the annual conference of the National Organization for the Professional Advancement of Black Chemists and Chemical Engineers	For a number of years, NIST has sponsored a symposium at the National Organization for the Professional Advancement of Black Chemists and Chemical Engineers to provide an overview of the agency, give presentations, have in-depth conversations about research areas in which its staff are engaged, and discuss its academic programs. NIST also has a booth to share information about its academic funding opportunities and to highlight the research activities of its Black scientific staff.
National Society of Black Physicists annual meeting	NIST is a co-sponsor of this meeting and has begun hosting a job opportunity booth.
Richard Tapia Celebration of Diversity in Computing (CMD-IT, 2021)	NIST hosts a booth where staff share information with students and educators about current NIST research and projects.
Society for Advancement of Chicanos/ Hispanics and Native Americans in Science	NIST participates in programs offered by the Society for Advancement of Chicanos/Hispanics and Native Americans in Science and shares information about current NIST research opportunities (Society for Advancement of Chicanos/Hispanics & Native Americans in Science, n.d.).
STEMversity	NIST research scientists participated for five years as workshop leaders and mentors for the summer academy of STEMversity, a nonprofit providing mentoring in forensic science for middle school and high school students in rural Georgia.
Textio	NIST will be piloting Textio, an augmented writing tool that helps companies craft more compelling and inclusive job advertisements, for all NIST job announcements in FY21 and FY22.
YWCA Coding Jam Session for Young Girls of Color	In October 2020 and February 2021 NIST hosted virtual coding jam sessions with YWCA Boulder County in Boulder, Colorado, targeted to middle school girls of color, encouraging more than 30 girls to pursue coding as a fun, promising career.

SOURCE: NIST leadership and employees, conversation with study director on behalf of the committee.

portant, it is also important that agencies consider structural inequities that may be present in their core programs and activities. Agencies should take steps to promote diversity, equity, and inclusion and mitigate biases in all aspects of their operations—from the recruitment, retention, and advancement of their workforce; to the processes that guide grant making and proposal review; to the scientific and technical areas that receive the most funding. Such efforts should be guided by qualitative and quantitative data collection and monitoring (NASEM, 2020).

THE IMPORTANCE OF ACCOUNTABILITY
FOR PROMOTING CHANGE

Despite efforts by Congress and federal government agencies to try to address the underrepresentation of specific groups in tech, there have not been significant improvements in the number of women of color entering, remaining, and advancing in tech fields (see Chapter 2). The widespread disparities that persist in tech education and careers underscore the need for additional efforts by a range of stakeholders to address structural racism and sexism in tech using an intersectional approach that makes use of the research on what is effective in driving change. Specifically, research demonstrates that increasing transparency and accountability in diversity, equity, and inclusion efforts can yield tangible positive results. Accountability, defined as "an obligation or willingness to accept responsibility or to account for one's actions," can promote individual and organizational-level behavior change (Merriam-Webster, n.d.; NSF, 2018, 2020a).

Recent studies by the National Academies and other groups to identify promising practices for promoting greater equity, diversity, and inclusion in STEM have coalesced around the importance of transparency and accountability for promoting change. For example, the 2018 report *Sexual Harassment of Women: Climate, Culture, and Consequences* emphasizes the importance of accountability in preventing and addressing sexual harassment in science, engineering, and medicine (NASEM, 2018). The authoring committee wrote: "One central, and perhaps more obvious, way to prevent sexual harassment is for academic institutions to clearly demonstrate that they do not tolerate it. . . . Doing so requires making the community aware that perpetrators of harassment are being held accountable and that the institution takes the matter seriously" (p. 143). The report offers specific examples of organizational accountability, such as this example from the NASA:

> An example of how organizations can hold leaders accountable can be seen in the policies and procedures used by NASA. Within NASA, managers and supervisors are considered not only as receivers and decision makers on allegations of harassment, but also as leaders who take action to prevent harassment in the workplace, and are accountable under the agency's annual performance

review system. Additionally, NASA produces an annual report on the functioning of its anti-harassment processes, which includes information on the number of cases addressed, the basis for each case (including sexual or nonsexual), the time required to process the case, and the remedial actions taken. This reporting process provides a mechanism for the leadership to monitor how the anti-harassment processes are functioning and whether changes or corrections need to be made (p. 149).

Similarly, the 2020 National Academies report *Promising Practices for Addressing the Underrepresentation of Women in Science, Engineering, and Medicine: Opening Doors* describes the research on the impact of transparency and accountability. The authoring committee wrote that "institutions must articulate and deliver on measurable goals and benchmarks that are regularly monitored and publicly reported. Multiple studies have demonstrated that transparency and accountability can drive behavior change" (Castilla, 2015; Dobbin and Kalev, 2016; Gaventa and McGee, 2013; Kruglanski and Freund, 1983; NASEM, 2020).

Accountability in equity, diversity, and inclusion efforts can take many forms. For example, *Promising Practices* describes several case studies:

Take, for example, Emilio Castilla's field study of the Massachusetts Institute of Technology's Sloan School of Management, where African Americans were consistently given smaller raises than white employees, despite identical job titles and performance ratings. To address this pervasive issue, Sloan began posting the average performance reviews and associated raises for each unit by demographic characteristics (i.e., race and gender). As soon as managers realized that bias in compensation by race and gender would become public knowledge within the school, they developed an increased sense of accountability and the discrepancies in compensation disappeared. Deloitte offers another compelling example. In 1992, chief executive officer Mike Cook realized that despite gender parity in hiring, the company was struggling to retain and advance talented women. He assembled a high-profile task force to address the issues with retention. Adopting a strategy that relied on accountability, the task force got each office within the company to monitor the career progress of its women and set goals to address the problem within the context of the specific unit (Castilla, 2015).

The report also described the positive impact of transparency and accountability on education, explaining that

[w]hen teachers realize that they will have to explain their evaluations, they rely less on their biases. For instance, studies have shown that when teachers are told that they will have to discuss and justify the grades they give students on papers, racial bias in grading disappears (Kruglanski and Freund, 1983). Equally, when departments are expected to present short lists of potential candidates to the

dean's office for review, those lists include more diverse candidates (Bilimoria and Buch, 2010; NASEM, p. 133).

Through the workshops held to inform the report, the committee also heard from leadership of the Department of Energy (DOE) about efforts to promote transparency, accountability, and data collection at the national laboratories in an effort by the DOE Office of Science to better understand how the national laboratories are working to foster diversity, equity, and inclusion in STEM. Modeled off an existing best practice for reporting 5-year strategic plans for science and technology strategies to senior leadership, DOE established a process for annually reviewing diversity, equity, and inclusion efforts by the national laboratories. In October 2016, the Office of Science issued a letter to the laboratories outlining the steps the agency would take to provide guidance on how to communicate these strategies and how the laboratories would be provided with annual feedback. This letter also instituted a requirement that laboratories would begin posting and updating their demographic data on their public websites annually. The laboratories were asked by DOE to address how they were assessing their diversity, equity, and inclusion (DEI) challenges, what their DEI goals were, what the roles and responsibilities of leadership and staff would be, and what their measures of progress and accomplishments were. The laboratories were also asked to include demographic data for their annual workforce and new hires.

Since 2017, DOE has been reviewing these strategies and providing feedback. In 2019, an external review panel—which included leaders in DEI at academic institutions, DEI leaders from scientific professional fields, and social scientists—was brought in to provide additional feedback. Laboratories were required to address the findings and recommendations of this panel as part of their FY2020 performance evaluation. The peer review identified promising practices under way at the national laboratories, including

- Efforts to incorporate DEI goals into performance evaluations of their leadership;
- Efforts to mitigate bias in recruitment and hiring;
- Family-friendly policies that address the needs of individuals at all stages of their careers;
- Mentorship during onboarding;
- Professional development opportunities to continue skills development;
- Laboratory-wide climate surveys to assess laboratory cultures and identify challenges; and
- Periodic surveys to assess progress.

The peer review also identified several areas in need of greater attention by national laboratories, which included a greater emphasis on data disaggregation. The DOE website currently posts demographic data for each job category;

however, the lack of data disaggregation presents a challenge in terms of tracking and understanding the experiences, recruitment, retention, and advancement of women of color in technology and other STEM fields.

The importance of transparency and accountability in diversity, equity, and inclusion efforts is becoming more widely recognized by researchers and members of the public alike. The section below describes efforts by investors to hold tech companies accountable for progress on their stated diversity goals and makes the case for the role of government in promoting transparency and accountability among tech companies, especially those that are recipients of government contracts.

A Role for Congress in Holding Tech Companies Accountable Through Transparently Sharing Workforce Demographic Data

Recognizing the importance of data collection, transparency, and accountability, many investors have called upon tech companies (and other large companies) to be more transparent about the makeup of their workforce. While companies consistently state their commitment to diversity, equity, and inclusion, and most have programs and initiatives intended to support progress in these areas (see Chapter 4), investors are calling for transparent, standardized data reporting so they can compare companies and hold them accountable. Specifically, many investors have called upon companies to make public the EEO-1 form (also known as a Standard Form 100), which most companies must submit to the Equal Employment Opportunity Commission (EEOC) annually in order to be considered an equal employment opportunity employer under Title VII of the Civil Rights Act of 1964 (as amended by the Equal Employment Opportunity Act of 1972) (U.S. Equal Employment Opportunity Commission, n.d.).[7] The EEO-1 form reports data on race, ethnicity, gender, and job category. The data are provided to the EEOC, and if an employer is also a federal contractor, the EEOC provides data obtained from the EEO-1 to the Office of Federal Contract Compliance Programs at the Department of Labor. The EEOC uses the data from the EEO-1 forms to support civil rights enforcement. It also uses the data to analyze

---

[7] U.S. Equal Employment Opportunity Commission. n.d. Title VII of the civil rights act of 1964. https://www.eeoc.gov/statutes/title-vii-civil-rights-act-1964. Companies required to submit an EEO-1 form to the EEOC are any private employer that has 100 or more employees (excluding state and local governments, public primary and secondary school systems, institutions of higher education, American Indian or Alaska Native tribes, and tax-exempt private membership clubs other than labor organizations) or any private employer that is subject to title VII and has fewer than 100 employees but is owned, affiliated with, or controlled by a company with more than 100 employees overall. Also required to submit an EEO-1 form to the EEOC are federal contractors with 50 or more employees that are prime contractors or first-tier subcontractors and have a federal government contract, subcontract, or purchase order amounting to $50,000 or more or serve as a depository of government funds in any amount or act as an issuing and paying agent for U.S. savings bonds and savings. See https://www.eeoc.gov/employers/eeo-data-collections.

employment patterns and to select certain employers for compliance evaluations. Both the EEOC and Office of Federal Contract Compliance Programs use statistical assessment of EEO-1 data to identify companies with indicators of systematic discrimination. Under Title VII, EEOC is required to keep the EEO-1 forms from individual companies confidential; however, the Office of Federal Contract Compliance Programs is not subject to these same requirements under Title VII.

Some investors believe that making the EEO-1 data publicly available will promote competition among companies that could drive companies to make greater progress on their stated goals related to workforce diversity, equity, and inclusion. In 2019, Intel became the first major tech company to publicly release its EEO-1 form (McGregor, 2019). The chief diversity officer of Intel, Barbara Whye, wrote that "transparency and open sharing of our data enable us to both celebrate our progress and confront our setbacks on that journey. We feel a sense of responsibility to continue to lead the industry in this space by raising the transparency bar for ourselves and, as a result, raising it for others. . . . Hopefully, openly sharing the details of our representation journey will encourage others in the industry to do the same" (Whye, 2019). Unfortunately, few tech companies readily followed Intel's example (Double Union, 2017; McGregor, 2019).

Nevertheless, calls by investors for companies to commit to public disclosure of data on the diversity of their employees have grown in the aftermath of the deaths of George Floyd, Breonna Taylor, and many others at the hands of police, and in July 2020 Calvert Research and Management, an investment firm active in encouraging companies to publicly disclose data, wrote to board chairs of the largest 100 U.S. companies by market value, asking them to release EEO-1 diversity data (Norton, 2020). As of 2021, half of the 100 largest publicly traded companies in the United States had agreed to publicly share their EEO-1 data, including the tech giants Alphabet, Amazon, Cisco Systems, and Salesforce.

Still, many tech companies that are the recipients of large government contracts have not publicly released their EEO-1 data. In its research, the committee found that of the companies[8] that were recipients of the 10 largest government contracts in tech in 2020 (which ranged in amount from $8,075,048,000 to $3,913,263,000; Washington Technology, 2020), not a single one had, at the time of publication of this report, released its EEO-1 data publicly.

In the committee's view, the public should be afforded the opportunity to hold these government contractors—the recipients of billions of taxpayer dollars—accountable for making progress toward their stated missions to improve the diversity of their workforce. The research on the impact of transparency and accountability on equity, diversity, and inclusion efforts strongly suggests that public release of the EEO-1 forms by the government could yield tangible, positive results. At the end of this chapter, the committee offers a recommendation for how Congress could

---

[8] This includes companies that provide information technology, telecommunications, consulting, professional, engineering, and other technology-driven products and services.

work to hold tech companies accountable through requiring release of EEO-1 workforce demographic data (see Recommendation 5-3).

Role of Federal Agencies in Incentivizing Greater Accountability

Federal agencies can also play a powerful role in incentivizing action at institutions of higher education through supporting programs that encourage transparent data collection and goal setting for efforts to promote diversity, equity, and inclusion. For example, agencies like NSF and the National Institutes of Health (NIH) are contributing to incentivizing efforts by institutions to improve diversity, equity, and inclusion in STEM by supporting the STEMM (science, technology, engineering, mathematics, and medicine) Equity Achievement (SEA) Change effort, an initiative led by the American Association for the Advancement of Science modeled after the Athena SWAN (Scientific Women's Academic Network) Charter in the United Kingdom. The Athena SWAN Charter was established in 2005 with a goal to "encourage and recognise commitment to advancing the careers of women in science, technology, engineering, maths and medicine employment." It has since expanded beyond the United Kingdom and been adopted by Ireland (Athena SWAN Ireland), Australia (SAGE-Athena SWAN), Canada (DIMENSIONS), and the United States (SEA Change).

The Athena SWAN framework has multiple components, including an award program through which institutions can gain recognition as a gold, silver, or bronze awardee, depending on the stage and success of their efforts to promote gender equity and representation. Though evaluating the direct impact of Athena SWAN on women in STEM is challenging because it is difficult to attribute cause and effect when other nation-wide and institution-wide efforts are simultaneously under way, one evaluation of the charter reported the perception among STEM professionals "of a positive effect of Athena SWAN on the visibility, leadership skills, career development, and satisfaction of women working in STEM and medicine, as well as the value of Athena SWAN as a driver in improving gender diversity" (Rosser et al., 2019).

The SEA Change effort in the United States resembles the Athena SWAN Charter in its emphasis on accountability through data collection, community support, educational resources, and an award system; however, it distinguished itself by its explicit focus on intersectionality. It is also notable that SEA Change was developed in an interdisciplinary manner that considered, among other things, the legal landscape and how it interfaces with efforts to promote diversity, equity, and inclusion. See Box 5-2 for additional description of SEA Change and Athena SWAN.

In addition to supporting specific programs, such as SEA Change, that incentivize progress through data collection and accountability, the committee believes that federal agencies should take additional steps to hold individual grantee institutions accountable for their stated goals to support diversity, equity,

BOX 5-2
The SEA Change Effort at the American Association for the Advancement of Science: Supporting Institutional Transformation in Support of Diversity, Equity, and Inclusion

Modeled after the Athena SWAN (Scientific Women's Academic Network) Charter (Advance HE, n.d.) in the United Kingdom, the STEMM Equity Achievement (SEA) Change effort (AAAS, 2021), launched by the American Association for the Advancement of Science in 2018, provides institutions with a community of practice; a range of educational resources; opportunities on best and promising practices for promoting equity, diversity, and inclusion; and positive incentives to work toward systemic change through an award program that recognizes institutions that undergo self-assessment, take action, and reassess in both a top-down and bottom-up manner.

The Three Pillars of the SEA Change effort include

- **Community**: SEA Change provides a "safe space where partnerships and collaborations can be established to nurture the talent pool for STEMM among member institutions, organizations, and individuals committed to advancing diversity, equity, and inclusion."
- **Institute**: The SEA Change Institute offers participating institutions with a repository of research; access to issue-based convenings, courses, trainings, and recordings of past SEA Change events; and information on how to apply for a SEA Change award.
- **Awards**: Participating institutions can apply for recognition by the SEA Change program for a bronze, silver, or gold award that recognizes "commitment to and creation of sustainable systemic change through self-assessment."

SEA Change resembles Athena SWAN in its emphasis on (1) establishing a community committed to principles of equity, diversity, and inclusion, and (2) a cycle of self-assessment followed by the adoption of evidence-based practices and the establishment of an action plan including reassessment and monitoring progress toward ambitious, but attainable, goals. Athena SWAN reports that over 100 institutions and 700 departments in the United Kingdom are engaged with the charter. SEA Change, while similar, differs in several important ways. Chief among them is that SEA Change places a much greater emphasis on race and ethnicity and the intersectional experiences of women of color, while Athena SWAN is focused primarily on gender. Another difference is that, in 2015, the Athena SWAN Charter was expanded to include "work undertaken in arts, humanities, social sciences, business, and law; in professional and support roles; and for transgender staff and students," while the SEA Change effort is currently focused on STEMM.

Box adapted from NASEM (2020).

and inclusion. This committee is not alone in expressing this view. The 2020 National Academies report *Promising Practices for Addressing the Underrepresentation of Women in Science, Engineering, and Medicine: Opening Doors* made the same observation and offered a set of specific recommendations related to the role of government in promoting greater accountability, which emphasized an intersectional approach (NASEM, 2020, recommendations 6-1 and 6-2, pp. 149-150). If these recommendations are implemented with the intentional focus on intersectionality, it is this committee's opinion that they could be a positive force for holding federal agencies and their grantees accountable for working in good faith to address the underrepresentation of women of color in tech education and careers. Building on these recommendations as a foundation, the committee offers a recommendation with a series of implementation actions at the end of this chapter focused on promoting accountability (see Recommendation 5-4 A-D).

SPHERE OF INFLUENCE AND THE IMPACT OF STRATEGIC PARTNERSHIPS ACROSS SECTORS

In this committee's experience, strategic partnerships that extend an organization's sphere of influence are key to promoting policy change. There are examples in science and education policy more generally in which meaningful policy change has grown out of partnerships and a coordinated advocacy effort. Advocacy coalition frameworks and specific case study examples could serve as models to stakeholders, such as scientific and engineering professional societies, that are working to advocate for improving the recruitment, retention, and advancement of women of color in tech (Weber, 2019; Weible, 2017; Weible and Ingold, 2018).

Take the example of climate change policy. While we are yet to have comprehensive climate change legislation at the national level, partnerships between scientists and environmental groups have played a key role in promoting legislation on renewable energy provisions and energy efficiency standards. Another example can be found in California's menu-labeling policy, enacted in response to the obesity epidemic, through which an advocacy coalition influenced a state health policy (Payán et al., 2017). Also, the recently passed FUTURE Act—which permanently extends mandatory funding to minority-serving institutions—had a coalition of stakeholders advocating for it, including representatives from more than 40 associations, members of which wrote 62,000 letters and made 3,000 phone calls to members of Congress (Long, 2019).

Furthermore, coalitions of professional groups with similar mission and scope can be an influential source of advice and guidance for government agencies. For example, the American Indian Science and Engineering Society, the Computing Alliance of Hispanic-Serving Institutions, and the United Negro College Fund, with funding from NSF, organized national convenings centered on discussions of how to increase representation of faculty from underrepresented groups and of minority-serving institutions in the NSF Computer and Information Science and Engineering Directorate's portfolio. The American Society for Engineering Education brought together faculty from Historically Black Colleges and Universities, Hispanic-Serving Institutions, Tribal Colleges and Universities, and other minority-serving institutions to amplify recommendations from these groups to NSF. NSF leadership and program directors in the Computer and Information Science and Engineering Directorate responded with the launch of a new, focused program to increase the number of minority-serving institutions and faculty from underrepresented groups who receive funding from its core programs. The Computer and Information Science and Engineering Directorate's Minority-Serving Institutions Research Expansion Program exemplifies the importance of elevating the visibility of organizations that have the knowledge and experiences with underrepresented groups to identify actions that can lead to meaningful change.

With these examples in mind, the committee sees an opportunity for scientific and engineering professional societies (e.g., American Association for the Advancement of Science, American Physical Society, American Chemical Society, National Society of Black Engineers) and higher education associations (e.g., the Association of American Universities) that engage in advocacy for science and for diversity, equity, and inclusion in STEM, to form strategic partnerships with influential organizations that have worked for many years to address structural racism and sexism and which have a great deal of influence with government institutions. For example, scientific and engineering professional societies could forge strategic partnerships with the NAACP, National Urban League, LULAC, UnidosUS, Native American Rights Fund, United Negro College Fund, the National Congress of American Indians, and other groups to expand their sphere of influence.

This committee also believes that the scientific community and individual scientists have an opportunity to become more engaged in the policy process as it pertains to the promotion of equity, diversity, and inclusion in STEM fields and in tech in particular. It has been this committee's observation that members of the scientific and engineering community tend not to engage often with their elected officials and seem, in general, to have an incomplete understanding of the policy making and appropriations process at the national level. Nor do members of the scientific community tend to engage with the state-level district offices of their elected officials (which does not require travel to Washington, DC).

In this committee's opinion, the scientific community should be engaging more actively with policy makers, especially around issues of diversity, equity, and inclusion in tech. Further, organizations that advocate on behalf of the scientific community should consider partnering with organizations that work to address structural racism to expand their sphere of influence on the issue of diversity, equity, and inclusion in STEM, with a particular emphasis on tech (see Recommendation 5-5).

RECOMMENDATIONS

The committee offers the following set of recommendations based on the information presented throughout this chapter related to the role of government in addressing the underrepresentation of women of color in tech.

RECOMMENDATION 5-1. Government efforts aimed at addressing the underrepresentation of particular groups in tech should intentionally account for intersectionality.

5-1 A. Any legislation aimed at addressing issues of underrepresentation in STEM and in tech should take an intersectional approach that considers the unique experiences of women of multiple marginalized identities (as described in Box 5-1).

5-1 B. Government efforts calling for data collection related to groups underrepresented in STEM and in tech should clearly indicate that such data be disaggregated by race/ethnicity and gender (to the extent possible given the need to protect anonymity of individuals) and should require qualitative as well as quantitative data collection, especially when the numbers are small enough that qualitative data would provide more meaningful information.

5-1 C. Program solicitations and descriptions at federal agencies should be explicit in directing prospective grantees to take an intersectional approach.

History demonstrates that unless policies, practices, programs, and individuals embrace an intersectional approach in efforts to promote diversity, equity, and inclusion in our institutions, women of color will likely continue to fail to fully benefit from these efforts. The committee found that both legislative language and program solicitations at federal agencies related to diversity, equity, and inclusion are inconsistent in calling for an intersectional approach.

RECOMMENDATION 5-2. Federal agencies should submit to Congress an overview of their programs that support the recruitment, retention, and advancement of women of color in tech with their annual budget request, as NSF currently does in its Summary Table on Programs to Broaden Participation (see Table 5-1). If agencies do not create such annual reports voluntarily, Congress should mandate that agencies do so.

In general, information about existing federal efforts aimed at supporting women of color in tech is widely dispersed and inconsistently distributed on various agency websites. The highly distributed nature of this information makes it challenging to gain a complete understanding and an accurate record of these investments. The one exception is NSF, whose annual budget request to Congress provides a compilation of the agency's efforts each year to support broadening participation.

RECOMMENDATION 5-3: To promote transparency and account-ability, Congress should amend section 709e of the Civil Rights Act of 1964 to require public release of EEO-1 workforce demographic data by companies, which would include those that are the recipients of government contracts supported by taxpayer dollars.

Research demonstrates that increasing transparency and accountability in diversity, equity, and inclusion efforts can yield tangible positive impacts. Recognizing the importance of data collection, transparency, and accountability, many investors have called upon tech companies (many of which are recipients of large government contracts) to be more transparent about the makeup of their workforce by publicly releasing the EEO-1 demographic data that most companies are required to provide to the Equal Employment Opportunity Commission annually. In the committee's view, the public should be afforded the opportunity to hold these government contractors—some of which are the recipients of billions of taxpayer dollars—accountable for making progress toward their stated missions to improve the diversity of their workforce.

RECOMMENDATION 5-4. Federal agencies should incentivize grantee institutions' efforts to improve diversity, equity, and inclusion through accountability measures.

5-4 A. Prospective grantees' plans to promote diversity, equity, and inclusion should be reviewed by review panels and agency personnel and should be a determining factor in awarding or renewing funding to an institution, in addition to technical merit. Grantees should include a description of the impact of their efforts to promote diversity, equity, and inclusion in annual reports and requests for funding renewals.

5-4 B. Federal agencies should invest in programs that incentivize institutional efforts to take a culturally responsive, intersectional approach in promoting diversity, equity, and inclusion in tech through award and recognition programs, such as the SEA Change effort led by the American Association of the Advancement of Science, which is currently funded by the National Science Foundation, the National Institutes of Health, and a number of private foundations.

5-4 C. Federal agencies should carry out periodic "equity audits" for grantee institutions to ensure that the institution is working in good faith to take an intersectional approach to address gender and racial disparities in recruitment, retention, and advancement.

- Institutions could be electronically flagged by the funding agency for an equity audit after a certain length of time or amount of funding is reached.
- An evaluation of the representation of women of color among leadership and academic success of women of color disaggregated by department should be included in such an audit.
- Equity audits should include a statement from institutions to account for the particular institutional context, geography, resource limitations, and mission and hold that institution accountable within this context. The statement should also account for progress over time in improving the representation and experiences of underrepresented groups in science, engineering, and medicine and should indicate remedial or other planned actions to improve the findings of the audit.
- The equity audit should result in a public-facing report made available on the agency's website.[9]

5-4 D. Federal agencies should consider institutional and individual researchers' efforts to support greater equity, diversity, and inclusion as part of the proposal compliance, review, and award process. To reduce additional administrative burdens, agencies could work within existing proposal requirements to accomplish this goal. For example, NSF could revise the guidance to grantees on its broader impact statements and the National Science Board could carry out a review of past NSF awards to determine how the NSF directorates have accounted for gender equity, diversity, and inclusion among the metrics evaluated in proposals submitted to NSF.

Federal agencies can play a powerful role in holding grantees accountable and by incentivizing action at institutions. If these recommendations are implemented with the intentional focus on intersectionality, it is this committee's opinion that they could be a positive force for holding institutions accountable for working in good faith to address the underrepresentation of women of color in tech education and careers.

> **RECOMMENDATION 5-5. Professional organizations and associations that represent the scientific and tech community (e.g., the Association for Computing Machinery, the Association for Computing Machinery, the Institute of Electrical and Electronics Engineers, the American Association for the Advancement of Science) should consider partnering with organizations that are committed to dis-**

---

[9] This recommendation is also put forth in NASEM (2020).

mantling structural racism, such as the NAACP, National Urban League, LULAC, UnidosUS, Native American Rights Fund, United Negro College Fund, and National Congress of American Indians, to extend their sphere of influence and expand their outreach to policy makers on issues related to diversity, equity, and inclusion in tech fields.

Strategic partnerships that extend an organization's sphere of influence are key to promoting policy change. There are examples in science and education policy in which meaningful policy change has grown out of partnerships and a coordinated advocacy effort. Advocacy coalition frameworks and specific case study examples could serve as models to stakeholders, such as scientific and engineering professional societies, that are working to advocate for improving the recruitment, retention, and advancement of women of color in tech (Weber, 2019; Weible, 2017; Weible and Ingold, 2018). The committee sees an opportunity for scientific and engineering professional societies (e.g., American Association for the Advancement of Science, American Physical Society, American Chemical Society, National Society of Black Engineers) and higher education associations (e.g., the Association of American Universities) that engage in advocacy for science and for diversity, equity, and inclusion in STEM, to form strategic partnerships with influential organizations that have worked for many years to address structural racism and sexism and which have a great deal of influence with government institutions.

REFERENCES

Advance HE. n.d. Athena SWAN Charter. https://www.ecu.ac.uk/equality-charters/athena-swan/.

American Association for the Advancement of Science. 2021. SEA Change. https://seachange.aaas.org.

Bailey, M. 2020. Between two worlds: Black women and the fight for voting rights. National Park Service. https://www.nps.gov/articles/black-women-and-the-fight-for-voting-rights.htm.

Bilimoria, D., and K. Buch. 2010. The search is on: Engendering faculty diversity through more effective search and recruitment. *Change: The Magazine of Higher Learning* 42:27-32. http://doi.org/10.1080/00091383.2010.489022.

Bleiweis, R., S. Phadke, and J. Frye. 2020. 100 years after the 19th Amendment, the fight for women's suffrage continues. August 18.

Castilla, E. 2015. Accounting for the gap: A firm study manipulating organizational accountability and transparency in pay decisions. *Organization Science* 26:311-333. https://doi.org/10.1287/orsc.2014.0950.

CMD-IT (Center for Minorities and People with Disabilities in IT). 2021. TAPIA Conferenc. https://tapiaconference.cmd-it.org/.

Crenshaw, K. 1989. Demarginalizing the intersection of race and sex: A black feminist critique of antidiscrimination doctrine, feminist theory and antiracist politics. *University of Chicago Legal Forum* 40(1):139-167.

Dobbin, F., and A. Kalev. 2016. Why diversity programs fail. *Harvard Business Review*. https://hbr.org/2016/07/why-diversity-programs-fail.

Double Union. 2017. Open diversity data. http://opendiversitydata.org/.

Gaventa, J., and R. McGee. 2013. The impact of transparency and accountability initiatives. *Development Policy Review* 31:s3-s28. https://doi.org/10.1111/dpr.12017.

Hammonds, E. M. 2020. Enshrining equity in democracy. *Science* 369(6508):1147. https://doi.org/10.1126/science.abe3003.

Kruglanski, A., and T. Freund. 1983. The freezing and unfreezing of lay-inferences: Effects on impressional primacy, ethnic stereotyping, and numerical anchoring. *Journal of Experimental Social Psychology* 19:448-468.

Long. K. 2019. A once-in-a-generation outcome: The FUTURE act signed by president, becomes law. *UNCF.* https://uncf.org/news/a-once-in-a-generation-outcome-the-future-act-signed-by-president-becomes-law.

McGregor, J. 2019. Intel has publicly revealed pay data showing most top executives are white men. *The Washington Post.* https://www.washingtonpost.com/business/2019/12/11/intel-has-publicly-revealed-pay-data-showing-most-top-executives-are-white-men/.

Merriam-Webster. n.d. Accountability. https://www.merriam-webster.com/dictionary/accountability.

Mid-Atlantic Quantum Alliance. n.d. https://mqa.umd.edu/workforce-education.

NASEM (National Academies of Sciences, Engineering, and Medicine). 2018. *Sexual harassment of women: Climate, culture, and consequences in academic sciences, engineering, and medicine.* Washington, DC: The National Academies Press. https://doi.org/10.17226/24994.

NASEM. 2020. *Promising practices for addressing the underrepresentation of women in science, engineering, and medicine: Opening doors.* Washington, DC: The National Academies Press. https://doi.org/10.17226/25585.

Norton, L. 2020. 50 of 100 largest U.S. public companies agree to release key diversity data. *Barron's.* https://www.barrons.com/articles/largest-public-companies-to-release-diversity-data-51608204600.

NSF (National Science Foundation). 2018. SBE science of broadening participation (SBE SBP). https://beta.nsf.gov/funding/opportunities/sbe-science-broadening-participation-sbe-sbp.

NSF. 2019. Frequently asked questions (FAQs) for the 2019-2020 ADVANCE solicitation. https://www.nsf.gov/pubs/2019/nsf19043/nsf19043.jsp#q9.

NSF. 2020a. Computer science for all (CSforAll): research and RPPs). https://beta.nsf.gov/funding/opportunities/computer-science-all-csforall-research-and-rpps.

NSF. 2020b. FY 2021 budget request to congress: Programs to broaden participation. https://www.nsf.gov/about/budget/fy2021/pdf/13_fy2021.pdf.

NSF. 2021. Broadening participation in computing (BPC). https://www.nsf.gov/cise/bpc/.

NWBC (National Women's Business Council). 2020. Women's inclusion in small business innovation research & small business technology transfer programs. Rochester, NY: Dawnbreaker. https://cdn.www.nwbc.gov/wp-content/uploads/2020/08/11124006/Women-In-SBIR-Report_NWBC_Final_2020-08-07.pdf.

Payán D., L. Lewis, M. Cousineau, and M. Nichol. 2017. Advocacy coalitions involved in California's menu labeling policy debate: Exploring coalition structure, policy beliefs, resources, and strategies. *Social Science & Medicine* 177:78-86. https://doi.org/10.1016/j.socscimed.2017.01.036.

Rosser, S., S. Barnard, M. Carnes, and F. Munir. 2019. Athena SWAN and ADVANCE: Effectiveness and lessons learned. *Viewpoint* 393(10171):604-608. https://doi.org/10.1016/S0140-6736(18)33213-6.

Society for the Advancement of Chicanos/Hispanics & Native Americans in Science. n.d. Cultivate diversity in STEM education and fields. https://sacnas.org.

U.S. Department of Defense. 2021. Diversity & inclusion: Special emphasis programs. https://www.dodea.edu/Offices/DMEO/programs.cfm?cssearch=549212_2.

U.S. Department of Defense STEM. n.d. DoD STEM. https://dodstem.us/participate/opportunities/.

U.S. Equal Employment Opportunity Commission. n.d. Freedom of Information Act. 5 U.S.C. § 552. https://www.eeoc.gov/foia.

U.S. Equal Employment Opportunity Commission. n.d. EEO data collections. https://www.eeoc.gov/employers/eeo-data-collections.

U.S. Equal Employment Opportunity Commission. n.d. Title VII of the civil rights act of 1964. https://www.eeoc.gov/statutes/title-vii-civil-rights-act-1964.

U.S. House of Representatives Committee on Science, Space, and Technology. 2021. Press release: Chairwoman Johnson and Ranking Member Lucas introduce legislation to increase diversity in the stem workforce. https://science.house.gov/news/press-releases/chairwoman-johnson-and-ranking-member-lucas-introduce-legislation-to-increase-diversity-in-the-stem-workforce.

Washington Technology. 2020. Washington technology 2020 top 100. Federal Procurement Data System. https://washingtontechnology.com/toplists/top-100-lists/2020.aspx.

Weber, R. 2019. Keys to successful advocacy: The role and value of coalitions. Naylor: Association Advisor. https://www.naylor.com/associationadviser/successful-advocacy-coalitions/.

Weible, C. 2017. The advocacy coalition framework. International Public Policy Association. https://www.ippapublicpolicy.org/teaching-ressource/the-advocacy-coalition-framework/7.

Weible, C., and K. Ingold. 2018. Why advocacy coalitions matter and how to think about them. *Policy & Politics* 46(2):325-345. https://doi.org/10.1332/030557318X15230061739399.

Whye, B. 2019. Intel's continued commitment to transparency and equity at all levels. Intel. https://newsroom.intel.com/editorials/intels-continued-commitment-transparency-equity-all-levels/#gs.co5now.

6

Alternative Pathways for Women of Color in Tech and the Role of Professional Societies

As technology becomes more central to a wider range of professional, business, and scientific service sectors, the need for workers with computing competencies over the next decade will continue to grow at a rate higher than the growth of the overall labor force (NASEM, 2018). According to the Bureau of Labor Statistics, computer occupations are projected to grow significantly faster than the average occupation. Moreover, from 2020 to 2030, employment of computer and information research scientists is projected to grow 22 percent, much faster than the average for all occupations (Bureau of Labor Statistics, 2021).

The demands and needs for employees, in particular those who can write algorithms, manage data, and have cybersecurity expertise, continues to grow in all sectors of the workforce. In fact, as stated in the National Academies of Sciences, Engineering, and Medicine report on the growth of computer science undergraduate enrollments, 67 of the bachelor's degree holders in the computing workforce have degrees from outside computer science (NASEM, 2018).

In addition to entering the technology and computing workforce with a bachelor's degree, other pathways include employer-offered training; certification courses offered by two-year colleges, four-year colleges, and other organizations; training programs offered by community-based and non-profit organizations; apprenticeship and re-entry programs; and digital badging (Chapple, 2006; Kvasny and Chong, 2006). This chapter describes alternative pathways into tech for women of color to fill the burgeoning demand for tech talent, and the role of professional societies in supporting women of color.

Despite the growth in non-degreed, non-linear pathways into tech, participation rates for women of color remain low. Table 6-1 shows the percentage of women in tech-related fields within the population of individuals whose highest level of educational attainment is less than a bachelor's degree. The percentage of women of color who enter tech fields without a bachelor's degree (i.e., with less than a high school education, with a high school diploma only, or having completed some college) is lower than that for white women and markedly lower than that for men. For example, of the total number of individuals who enter tech-related fields with less than a high school diploma, 73.6 percent are men while only 26.3 percent are women and 15.2 percent are women of color. Overall, the percentage of individuals working in tech-related fields with less than a bachelor's degree is 41 percent.

ALTERNATIVE PATHWAYS INTO TECH

Non-linear pathways into tech careers are on the rise. In 2015, 51 percent of undergraduate credentials were awarded at the sub-baccalaureate level, compared to 48 percent in 2003 (Zhang and Oymak, 2018). These pathways could broaden opportunities for women and others who choose to pursue technology careers without the requisite college degrees that most traditional associate's and bachelor's degree programs offer. As Table 6-1 suggests, however, the percentage of women, and women of color specifically, who enter the fields at the sub-baccalaureate level is low.

Technology training programs are offered in multiple venues to prepare graduates to enter the tech workforce (Mardis et al., 2018). These programs, delivered by community (two-year) colleges, community-based organizations, and online by industry groups typically comprise four components: recruitment, support services, instruction, and job placement services (Shah, 2020). Grassroots or community-based programs that uniquely engage women of color and other underrepresented groups tailor their recruitment messaging, instruction, wraparound services (such as childcare and transportation), and ongoing support of their alumna in the workplace (Abbate, 2018).

As discussed in Chapter 2, many women of color have used the technology field as a means to escape poverty (Kvasny, 2006; NASEM, 2018). The alternative pathways described in the following subsections provide hope and inspire further education, thus presenting significant opportunities for women of color.

Community College and Career Technical Programs

High school career and technical education programs and community colleges offer guided pathways for preparation and skill building for entry into the tech industry through curricular frameworks, systems of digital badges for non-traditional student populations, and sub-baccalaureate credentials. Earned

TABLE 6-1 Racial/Ethnic Demographics of Individuals in Tech-Related Fields Without a Bachelor's Degree

Academic Attainment	Total Number (Women and Men)	Women								Men
		White (non-Hispanic)	Black or African American (non-Hispanic)	American Indian / Alaska Native (non-Hispanic)	Hispanic	Asian (non-Hispanic)	Native Hawaiian and Other Pacific Islander (non-Hispanic)	Identify as some other race (non-Hispanic)	Identify as Two or More Races (non-Hispanic)	All Racial/ Ethnic Groups
Less than high school	52,263	11.1%	4.5%	0.25%	6.8%	2.9%	0%	0.10%	0.62%	73.6%
High school graduate	351,456	20.4%	3.8%	0.22%	3.8%	0.60%	0.11%	0.15%	0.42%	70.7%
Some college	1,234,753	16.5%	3.3%	0.17%	2.0%	1.0%	0.02%	0.03%	0.59%	76.3%

SOURCE: 2018 U.S. Census Bureau's American Community Survey Public Use Microdata Sample files.

NOTE: The tech fields included are computer and information systems managers, computer and information research scientists, computer system analysts, computer programmers, information security analysts, software developers, software quality assurance analysts and testers, web developers, web and digital interface designers, computer support specialists, database administrators and architects, network and computer systems administrators, computer network architects, computer occupations, computer numerically controlled tool operators and programmers, and computer, automated teller, and office machine repairers.

sub-baccalaureate credentials include professional certifications or associate's degree programs below the bachelor's degree level.

Taking career and technical education courses (CTE) in high school is associated with a higher probability of graduating from high school and enrolling in a two-year college. Such courses increase a student's likelihood to be employed the year after graduation, resulting in a boost in wages. CTE courses, however, tend to be overrepresented by white male participants. Moreover, they lack curricula and stated learning outcomes to teach and assess critical "soft skills" that employers increasingly request.

A growing number of companies are collaborating with community colleges to align curricula to specific industry needs and trends in high-demand fields such as computer and information technology. Such partnerships ensure the graduates from the degree and certification programs acquire relevant career-ready skills, while providing employers with access to a future workforce. In addition, community and technical colleges that offer stipends for student participation mitigate financial obstacles that often provide a barrier for entry into workforce development.

A study of eight community and technical colleges in Washington state found that information technology programs were structured in four dimensions: (1) detailed program requirements listed required and elective courses, and programs utilized a cohort model and offering schedule flexibility; (2) program alignment was linked to industry needs and local employment opportunities; (3) students had ready access to information through the website and other informational resources; and (4) students had access to active advising and support that offered counseling for undecided students, group sessions, orientations, and support and monitoring of student progress (Van Noy et al., 2016) (Box 6-1).

Despite these multiple opportunities, as Table 6-1 illustrates, women and specifically women of color are not entering technology fields through these sub-baccalaureate pathways at the rates they choose to enter service-related industries such as education (Zhang and Oymak, 2018). Tribal colleges and universities and community-based training programs seek to redress this gap. With a focus on cultural relevance, tribal colleges and universities lead the nation in producing a highly educated and skilled Native workforce. Because of their mission to provide job training and technical and vocational education to prepare Native students for the workforce, tribal colleges and universities are well positioned to increase the number of Native women students in technology and computing; however, only 12 out of 35 tribal colleges and universities have computing programs (Tribal College Journal, 2019). The disparity is due in large part because of the lack of instructors.

BOX 6-1
A Model for Transitioning to College-Level Programs

Few adults, who have a high school education or less and who benefit from postsecondary occupational education and a credential, make the transition to college-level programs. To address this concern, Washington State's Integrated Basic Education and Skills Training (I-BEST)[1] program, designed by the Washington State Board for Community and Technical Colleges in conjunction with Washington's 29 community colleges and five technical colleges, used a combination of basic skills and professional technical instruction to allow students with basic skills to enter directly into college-level coursework. The goal of the I-BEST program is to increase rates of advancement to college-level occupational programs and rates of completion of postsecondary credentials in fields with opportunities for career advancement and good wages among English-as-a-second-language students and adult basic education students. Students in this program are generally more likely than the general population of basic skills students to be female, older, enrolled full time, and to have a GED or high school diploma. In addition, a higher proportion of these students are in the lowest socioeconomic status quintiles. A field study (Wachen et al., 2010) showed that I-BEST students achieved better educational outcomes than other basic skills students and were more likely to continue into credit-bearing coursework, earn college credits, attain occupational certificates, and make point gains on basic skills tests. A companion paper to Wachen and colleagues' (2010) report also found that I-BEST students had better outcomes than basic skills students who took at least one college-level occupational course and that the observed effects were causal and not merely correlational (Zeidenberg et al., 2010).

---

[1] More information about Washington's I-BEST program can be found at: https://www.sbctc.edu/resources/documents/about/facts-pubs/i-best.pdf.

Community-Based Programs

Community-based technology training programs provide instruction in supportive environments for tech-minded women to gain entry-level information technology skills for mostly low-wage jobs ($11 to $15 per hour). Programs provided by PerScholas, Npower, and G|Code, for example, tend toward providing practical hands-on training as opposed to teaching theoretical concepts in computer science and technology and distinct from those offered by community colleges, universities, or industry. Some programs, such as G|Code, offer residential housing for women who are selected for the program.

Community-based programs tend to be low or no cost and, therefore, appeal to people who are unable to afford to continue their traditional education after high school. G|Code and similar programs target women on the verge of aging out of homeless shelters and other temporary living situations. In addition to hands-on technical skills such as web development, coding, and cloud computing courses, the programs teach students about changes in technology and salary negotiation, to optimally position them as future employees and/or self-employed contractors. These programs and the certifications they confer can inspire and encourage the women to pursue further education (Carnevale et al., 2012).

Industry-Based Programs

Companies such as Apple, Cisco, Dell EMC, Google, IBM, Microsoft, Oracle, and VMware offer training programs to certify people in the technology that they produce.[1] For instance, Google's developer training provides free online courses in web development, machine learning, and Android development, the company's mobile device operating system.[2] The Android course in particular welcomes participants with a range of experience, from beginners to experienced developers. Those who complete the course have an opportunity to take a certification exam to acquire an additional credential they can leverage to secure a job and/or advance their career. Similarly, Apple offers training on its iOS and MacOS operating systems and Swift programming languages.[3] However, unlike Google, which has partnered with Udacity to offer free online courses, Apple leverages a network of authorized global training providers that can charge hundreds to thousands of dollars for the basic training courses, creating barriers to entry for low-income women of color.

IBM's Skills Academy training programs provide a pathway to earning an advanced degree. The training programs provide an educational benefit through a partnership with higher education institutions, including Northeastern University, the first academic institution to offer academic credit for digital badges earned through industry courses. IBM employees, customers, and members of the public can use IBM-issued badge credentials toward professional master's degree programs (Leaser et al, 2020). As will be discussed later in this chapter, digital badges provide visual indicators of performance that mark key learning accomplishments.

While industry-specific programs are proliferating, as the data in Table 6-1 suggest, women of color remain an untapped segment of their recruitment and enrollment strategies, and thus the future workforce.

---

[1] For more information see https://training.apple.com; https://www.cisco.com/c/en/us/training-events/training-certifications/certifications.html#~certifications; https://education.dellemc.com/content/emc/en-us/home/certification-overview.html; https://grow.google/programs/it-support; https://www.ibm.com/certify; https://www.oracle.com/news/announcement/oracle-offers-free-training-and-certification-for-oracle-cloud-infrastructure-2021-09-08/; and https://www.vmware.com/education-services/certification.html (accessed October 19, 2021).

[2] For more information see https://developers.google.com/training.

[3] For more information see https://training.apple.com/us/en/courses.

Apprenticeships and Re-entry Programs

In response to the growing need to expand the pool of prospective workers, university and business leaders have turned to apprenticeships and other "earn and learn" opportunities to provide job seekers with a pathway to earn postsecondary credentials and gain critical job experience and better wages. The apprenticeship programs particularly appeal to those who are out of work and have low levels of education and work experience, because they connect job seekers to both postsecondary education and a career (Beer, 2018) in more engaging ways than traditional schooling (Lerman, 2016).

Filling such a need is Apprenti, a national organization that bridges the tech talent and diversity gaps by adapting the apprenticeship model to meet evolving workforce needs.[4] Apprenti interviews and develops a custom course of training for participants, lasting two to four months, prior to placing them in the workforce. Some apprentices take coding classes full time and then start work, while others begin with a mix of classes and work. Upon successfully completing the technical training, apprentices continue for one year of paid on-the-job training. Apprenti partners with coding boot camps such as Code Fellows, Coding Dojo, Galvanize, TLG Learning, and others.

Twilio, a software company that enables phones, voice over IP (VoIP), and messaging to be embedded into web, desktop, and mobile software, offers the Hatch Apprenticeship Program, a six-month software engineering apprenticeship program for people from underrepresented groups.[5] The Hatch program is designed to provide access to software engineering roles for people with nontraditional educational, professional, and personal experience.

Despite these positive examples, apprenticeships, which for certain industries have been considered effective training vehicles, have yet to be widely adopted in information technology. Beer (2018) offers strategies for what could be applied to the development of a national apprenticeship system in the United States focused on information technology. Such a system would strive to (1) increase and align funding for postsecondary education and workforce development with emerging sectors; (2) increase the diversity of participants, with an intentional focus on engaging communities of color and women; and (3) expand access to pre-apprenticeships and youth apprenticeships aligned with postsecondary pathways (Beer, 2018).

Another specific opportunity that can increase the representation of women of color in the tech workforce opens a pathway for women who have taken career breaks and want to return to work. As many as 50 companies including Apple, IBM, Johnson and Johnson, and United Technologies have created "returnship" programs for women who want to specialize in tech jobs (Lipman, 2019). Operated like internships, and lasting from eight weeks to six months, returnships en-

---

[4] For more information see https://apprenticareers.org.
[5] See https://www.twilio.com.

able people returning to the tech workforce to refresh their skills, evaluate future employers, and be evaluated by future employers. Approximately 85 percent of those who participate in re-entry programs are hired permanently, according to iReLaunch, a career re-entry firm (Lipman, 2019).[6] A growing number of firms are offering opportunities for graduates to immediately enter the workforce. Though not specifically in tech, Ford's direct-hire program offers skills training and assigns a mentor and a "buddy" to help the re-entrants navigate the company's culture.

Re-entry programs provide an important opportunity for women who, after having raised their children, for example, are looking for new opportunities in the workplace that they previously may not have considered. The COVID-19 pandemic may also highlight an additional opportunity to develop re-entry and retraining programs to bring more women seeking opportunities in tech—many of whom left their jobs as a result of the pandemic—back into the workforce. However, it remains to be seen how large a potential pool of women of color are attracted to such opportunities, as they often have not had the luxury of remaining out of the workforce to raise their children in the first place.

Certifications

Most information technology training programs, whether community or industry based, prepare students to pass certification exams that demonstrate the knowledge, skills, and abilities that are recognized and desired by industry. The content of certificate programs signals key credentials in people that employers are seeking to hire or to promote from within (Carnevale et al., 2012). Georgetown University Center on Education and the Workforce has suggested that successful certification programs promote both gainful employment and the pursuit of a higher education degree credential, such as an associate's or bachelor's degree (Carnevale et al., 2012).

The most popular certificate program in technology and computing is the internationally recognized A+ Certification issued by the non-profit Computing Technology Industry Association (CompTIA),[7] a trade association that issues professional certifications[8] for the information technology industry. The A+ certification demonstrates vendor-neutral competency as a computer technician. CompTIA also certifies progressively more advanced skills including Cloud+ (cloud computing and virtualization), Network+ (design, configure, and manage wired and wireless devices), Server+ (server-specific hardware and operating environments), and Security+ (information technology network and operational security). The advanced security professional certificate issued by CompTIA is an example of a master level certification that is intended to build on the Security+

---

[6] See https://www.irelaunch.com/paidcorporateprograms.

[7] See https://www.comptia.org/home.

[8] See https://en.wikipedia.org/wiki/CompTIA.

credential. Holders of this certificate demonstrate the technical knowledge and skills required to "conceptualize, design, and engineer secure solutions across complex enterprise environments."

As noted previously, industry offers another pathway to demonstrate competency in their own technology platforms. For example, Google's Developers Certification is an imprimatur for professional Android developer, an associate cloud engineer, or a professional data engineer, among others. The company offers certificates for data analysts, project managers, UX developers, and information technology support specialists, and provides applicants with the median annual wage for each of the positions.[9] IBM's Skills Academy provides opportunities for students to become certified in artificial intelligence, cloud computing, the Internet of Things, blockchain, and other areas.[10]

There is a growing effort to share industry-recognized certifications to verify knowledge, skills, and abilities across states, and a mechanism to allow states to hold institutions accountable for credential attainment. The Certification Data Exchange Project[11] is one such effort (ACTE, 2017).

A significant opportunity exists to expand the capacity of community-based programs discussed earlier to prepare students, in particular, women of color, for both beginner and advanced certification exams. These programs' capacity to supplement traditional training programs with online resources needs to be expanded in order to provide progressive skills at a discounted price and widen the alternate pathways for women of color to pursue employment in tech that extends beyond entry-level positions. This hybrid model is being employed by NPower in eight major cities in the United States and Canada, although it is not clear from its stated mission or strategic plan whether there is a specific focus on women of color.

Women of color, however, are severely underrepresented in these multiple pathways and thus represent an untapped resource for talent, social exchange, and individual and community success. Figure 6-1 shows the number of certificates awarded in academic years from 2014-2015 to 2018-2019.

The number of certificates in computing shows considerable differences across gender and race/ethnicity as illustrated in Figure 6-1. The number of certificates awarded to white men and men of color increased between 2014 and 2018, while the number of certificates awarded to white women and women of color remained relatively flat. Women of color comprise the smallest segment of certificate awardees over the same period, suggesting structural barriers to this pathway. On average, women of color make up about 11 percent of the total of certificates awarded in computing in the last five years for all types of certificates earned (less than one academic year, at least one but less than two academic years, and at least two but less than four academic years). Figure 6-2 shows

---

[9] See https://grow.google/certificates/.

[10] See https://skills-academy.comprehend.ibm.com.

[11] See https://www.acteonline.org/certification-data-exchange-project/.

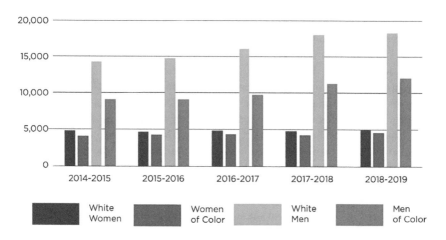

FIGURE 6-1 Number of certificates, which take less than four academic years, awarded in computing 2015-2019.
SOURCE: National Center for Education Statistics. Integrated Postsecondary Education Data System: 2014-2015 to 2018-2019, Certificates awarded in computing (CIP short-code 11) to women of color (Hispanic/Latino women, Black/African American women, Asian Women, American Indian/Alaska Native women, and Native Hawaiian/Pacific Islander women).

the types of certificates that are sought by women of color are by far those that require less than one academic year followed by those that require at least one year, but less than two academic years.

Certificate programs that were shorter in duration had higher participation from Black and Latinx women. During the 2018-2019 academic year, of the total number of certificates awarded to women of color in computing, Latinx women received 33 percent and Black women received 46 percent of certificates of less than one academic year awarded. For certificates requiring at least one but less than two academic years, Latinx women received 50 percent and Black women received 32 percent of certificates awarded. For certificates of at least two but less than four academic years, Latinx women received only 10 percent and Black women received just under 11 percent (NCES, 2020).

Badges and Gamification

Digital badges are increasingly being used to motivate students to increase engagement with coursework and reduce knowledge gaps (Besser and Newby, 2020; Gibson et al., 2015), and their effects on engagement do not vary by age, sex, or racial status (Higashi and Schunn, 2020). Such uses can be used to close equity gaps in tech.

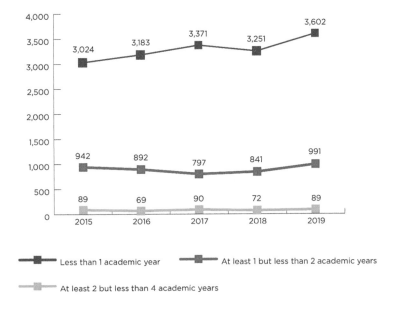

FIGURE 6-2 Comparison of certificates in computing awarded to women of color based on type of certificate 2015 to 2019.
SOURCE: National Center for Education Statistics. Integrated Postsecondary Education Data System: 2014-2015 to 2018-2019, Certificates in computing (CIP short-code 11) awarded to women of color (Hispanic/Latino women, Black/African American women, Asian Women, American Indian/Alaska Native women, and Native Hawaiian/Pacific Islander women) based on type of certificate (less than 1 academic year, at least 1 academic year, but less than 2 academic years, at least 2 academic years, but less than 4 academic years).

Badges provide visual indicators of performance that mark key learning accomplishments and scaffold the learning process by progressively moving students toward deeper understanding and learning independence (Devedžić and Jovanović, 2015). Badges also provide "taggable" ways in which employers can make decisions about prospective talent and provide teachers with critical feedback about their students' learning not available through traditional assessments and interactions (Moore, 2018; Wardrip et al., 2016).

Badges can be "gamified," allowing learners to compete with themselves and others with regard to specific learning outcomes (Gibson et al., 2015). Gamification of learning using digital badges can increase learners' extrinsic motivation and level of participation (Abramovich et al, 2013; Facey-Shaw et al., 2020). Moreover, the use of digital badges is associated with higher levels of confidence in technology integration skills, more courses taken, and higher overall course grades compared to traditional instructional methods (Newby and Cheng, 2020). Peer-awarded digital badges have also been effective at increasing engagement and fostering achievement (Rehak and Hickey, 2013). Thus, choosing what badges to include in curricula, and from whom they are awarded, must take into consideration both the motivation and skill levels of students. These benefits associated with badges suggest that they can be leveraged to motivate women of color (and others) to progressively increase their knowledge, skills, and abilities in tech.

Typically, by scaffolding the learning process, collections of achieved badges lead to certifications. The Carnegie Mellon Robotics Academy, in partnership with the University of Pittsburgh and the RAND Corporation, is studying the usefulness of badges to teach robotics to students of all ages.[12] Figure 6-3 depicts how badges are used to incentivize students to complete assignments (projects, assessments, portfolios, etc.) as they work toward certification. In the example shown in Figure 6-3, students must earn all four badges and receive approval from a certified instructor before they are allowed to take the Introduction to Programming Certification Exam.[13]

Open digital badges are web-enabled tokens that extend the motivational influences and credentialing of localized gamification to recognize skills and competencies acquired across formal or informal, online and traditional learning settings (Devedžić and Jovanović, 2015). This interoperability of badges across learning platforms is a natural evolution of the use of this learning tool, although best practice frameworks for successful implementation across institutions are still being formed (Voogt et al., 2016). The Extreme Science and Engineering Discovery Environment (XSEDE), an NSF-funded virtual organization, provides training opportunities to secure Open Badges in High-Performance Computing and other topics. Programs are offered for students at the beginner, intermediate, and advanced levels of competency who complete the OpenACC workshop. XSEDE awards Mozilla Open Badges to learners who demonstrate certain competencies (Kappes and Betro, 2015).

The federal government, private industry, and non-profits are working to standardize badging across digital platforms. For instance, the HASTAC/MacArthur Foundation Badges for Lifelong Learning Competition, in partnership with the Department of Education, seeks to elevate the value of specific learning outcomes, though the program is limited in size (Claussen, 2017).

Although digital badges have the potential to make learning pathways more visible and appealing to a wide range of audiences, there is little evidence in the literature that gamifying learning, interoperable platforms, and credentials have a specific effect on building confidence and motivating women of color to pursue information technology careers (Higashi and Schunn, 2020; Pitt et al., 2019). More research is needed to understand if badges encourage women of color to pursue learning and career opportunities in information technology, and what specific motivating mechanisms they can leverage.

---

[12] See https://www.cmu.edu/roboticsacademy/Research/Badges_cmra.html.

[13] See https://www.cmu.edu/roboticsacademy/Training/certificationsv2.html.

Badge Pathway to Certification

LEVEL 3 - Earn the Certification - Earn Badges + upload artifacts
+ get teacher endorsements + pass the final exam

FIGURE 6-3 Carnegie Mellon's Robotics Academy Badge Pathway to Certification.
SOURCE: Carnegie Mellon Robotics Academy at Carnegie Mellon University. See https://
www.cs2n.org/teachers/badges.

THE ROLE OF PROFESSIONAL SOCIETIES

Professional societies generally offer educational and informational resources and can offer various kinds of support for students who are interested in educational and career opportunities in a specific discipline (Morris and Washington, 2017; NAS, NAE, and IOM, 2005). Societies may publish professional journals, set standards, and help shape policy and curricula. Additionally, for those with programming, they may focus on particular groups, such as students, or a mix of students and professionals (NAS, NAE, and IOM, 2005). For the purposes of this report, professional societies are further defined as those organizations that operate as membership-based "societies" rather than community-based organizations and professional[14] or trade associations.

Professional societies generally support the development of standards and are positioned in a space where they can "design and promote change, including through publications, policy statements, meetings, committees, lectureships, and awards" (NAS, NAE, and IOM, 2005, p. 138). Unfortunately, this influence is not often exercised as effectively as it could or should be concerning increasing diversity. For example, in the design of processes for identifying and selecting recipients of awards and fellowships, women of color—particularly those who have entered tech through alternative or non-linear paths—may be disadvantaged in nomination processes that rely on networks where women of color have been historically excluded. These awards and fellowships are often used to demonstrate excellence and achievement as one works toward positions of leadership. Thus,

---

[14] The Computing Research Association (CRA) and AnitaB.org are examples of professional organizations that provide programming for women in tech.

when the processes for selecting recipients of awards and fellowships perpetuate a lack of diversity in potential recipients, they can create structural barriers that can limit the advancement of women of color in their careers (Ham, 2020; Hu, 2019). While professional societies may episodically ramp up their outreach to women and individuals from underrepresented groups, they often experience little success in actually increasing engagement or participation of such groups. The real key to engaging and broadening participation is designing programs and initiatives that are shaped by and for the groups they purport to target (Morris and Washington, 2017).

Recognizing this lack of available programming and support for people of color by traditional professional societies, and spurred by the civil rights, Chicano, American Indian, and women's movements, new organizations emerged in the 1960s and 1970s to serve those populations. Today, the same professional societies that primarily serve people of color and women are among the small number that have targeted programs or initiatives for women of color in tech. A review of the longstanding professional societies that support those in one or more of the STEM fields revealed no programs or initiatives focused specifically on creating pathways or advancing women of color in tech, though some do have programs or initiatives for women in tech and/or STEM, and/or people of color in tech and/or STEM. In addition, the leadership of professional societies most often does not reflect the diversity of the future workforce. Even among those professional societies that specifically serve people of color, there are very few with any programming or initiatives solely for women of color in tech. With this as a disappointing backdrop, the following societies are those that have programs or initiatives available to support women in tech and/or STEM, and/or people of color in tech and/or STEM.

American Association for the Advancement of Science

The American Association for the Advancement of Science (AAAS) is the world's largest multidisciplinary scientific society whose stated mission is to "advance science, engineering, and innovation throughout the world for the benefit of all people." AAAS has members in more than 91 countries and is a leading publisher of research through its *Science* family of journals. AAAS's STEM Equity Achievement (SEA) Change initiative supports diversity and inclusion, especially in colleges and universities by providing support to institutions as they work to transform systems and processes.[15] SEA Change provides resources (e.g., trainings, courses, published research), communities of practice for participating

---

[15] American Association for the Advancement of Science. (n.d.). Mission. Accessed October 31, 2020 from https://www.aaas.org/mission.

institutions and organizations, and awards to institutions that demonstrate commitment to sustainable change.

American Indian Science and Engineering Society

The American Indian Science and Engineering Society (AISES) is a national non-profit organization focused on substantially increasing the representation of American Indians, Alaska Natives, Native Hawaiians, Pacific Islanders, First Nations, and other Indigenous peoples of North America in STEM education and careers.[16] AISES partnered with the Women of Color in Computing Collaborative to expand computer science education for Native girls in Native-serving high schools. The project, "Expanding Computer Science Opportunities for Native Girls," aims to increase interest, engagement, and participation in computing education (with an emphasis on participation and success in advanced placement computer science courses) among Native high school students and girls/LGBTQ+/Two-Spirit students. The project is developing a series of courses from introductory level to advanced placement computer science courses, developed to include culturally relevant activities and modules aligned with tribal cultural values, vision, and goals for sovereignty.[17]

American Mathematical Society

Created in 1888, the stated mission of the American Mathematical Society (AMS) is to "further the interests of mathematical research and scholarship, and serve the national and international community through its publications, meetings, advocacy and other programs." AMS's membership of 30,000 individuals and 570 institutions supports the mathematical sciences by providing access to research, professional networking, conferences and events, advocacy, and a connection to a community passionate about mathematics and its relationship to other disciplines and everyday life. AMS sponsors programs and provides funding for educating and recruiting young people and people from underrepresented groups, helping them to develop a strong sense of belonging among the diverse global math community.[18]

---

[16] American Indian Science and Engineering Society. (n.d.-a). About. Accessed October 20, 2020 from https://www.aises.org/about.

[17] American Indian Science and Engineering Society. (n.d.). *Expanding Computer Science Opportunities for Native Girls*. Retrieved October 20, 2020, from https://www.aises.org/content/expanding-computer-science-opportunities-native-girls.

[18] American Mathematical Society. (n.d.). About the AMS. Retrieved October 31, 2020 from https://www.ams.org/about-us/about.

American Society of Engineering Education

The American Society of Engineering Education (ASEE) develops policies and programs that enhance professional opportunities for engineering faculty members, and promotes activities that support increased student enrollments in engineering and engineering technology in colleges and universities. ASEE provides a channel of communication between corporations, government agencies, and educational institutions. ASEE's organizational membership is composed of 400 engineering and engineering technology colleges and affiliates, more than 50 corporations, and numerous government agencies and professional associations.[19]

ASEE received several grants in the 1970s to research the status of women and American Indians and develop programs to attract more of these students to enter engineering. Since then, ASEE has continued to release studies on the subject in its *Journal of Engineering Education*, and has created divisions specifically devoted to developing programs and research in this area.[20] ASEE's Black Engineering College Development program used industry funding to support African American faculty or students at traditionally Black colleges in the pursuit of doctoral degrees in order with the goal of increasing the teaching workforce at historically Black colleges and universities. It has also conducted research on the status of women and American Indians in order to develop programs to encourage more of these students to pursue engineering and continue to develop programs and conduct and encourage research in this area.

Association for Computing Machinery

The Association for Computing Machinery's Women in Computing (ACM-W) special interest group supports, celebrates, and advocates internationally for the full engagement of women in all aspects of the computing field, providing a wide range of programs and services to association members and working in the larger community to advance the contributions of women in technical and computing fields. In addition to ACM-W, the Association for Computing Machinery (ACM) established a Diversity, Equity, and Inclusion Council in 2019 to serve as a convener for diversity, equity, and inclusion issues within ACM as well as a resource the ACM's special interest groups, conferences, boards, and councils looking for best practices. The ACM also established the Standing Committee on Systemic Change in 2020 whose goals include considering where changes can be made to promote racial equity in ACM volunteer activities, identifying and prioritizing opportunities for systemic change, addressing identified problems, and developing metrics for reporting diversity numbers.

---

[19] American Society of Engineering Education. (n.d.). About Us. Retrieved October 31, 2020 from https://www.asee.org/about-us/the-organization.

[20] American Society of Engineering Education. (n.d.-a). About Us-The Organization-Our History. Retrieved October 31, 2020 from https://www.asee.org/about-us/the-organization/our-history.

Institute of Electrical and Electronics Engineers Computer Society

The Institute of Electrical and Electronics Engineers (IEEE) Computer Society was created to be a source of information, inspiration, and collaboration in computer science and engineering. Its resources include international conferences, peer-reviewed publications, a robust digital library, globally recognized standards, and continuous learning opportunities.[21] The IEEE Computer Society offers resources, such as access to career development forums, to empower career development and salary potential of women in tech.[22]

National Society of Black Engineers

The mission of the National Society of Black Engineers (NSBE) is "to increase the number of culturally responsible Black engineers who excel academically, succeed professionally, and positively impact the community."[23]

The mission of NSBE's Women in Science and Engineering group is to enlighten, engage, and empower not only NSBE women in STEM but foster relationships and collaborate with communities and institutions outside of the society. NSBE also continues to build and establish Women in Science and Engineering as a foundational special interest group for both NSBE collegiate and professional members. The group hosts an annual roundtable discussion and webinar; promotes career development, holds workshops, and offers education grants for women engineers; aims to continue to increase its membership and aid in the retention of women engineers at institutions of higher education; aims to have a positive impact on communities, college campuses, and workplaces; provides mentorship opportunities for students transitioning from college to the workforce; provides a forum for women to provide mentorship opportunities for students transitioning from college to the workforce; and provides a forum to discuss topics and issues specific to gender- and STEM-related topics pertinent to Women in Science and Engineering (WISE).[24]

In addition to Women in Science and Engineering, which supports collegiate and professional women of color, NSBE hosts a number of pre-collegiate initiatives aimed in part to stimulate interest in engineering among girls of color. The Summer Engineering Experience for Kids (SEEK) program is a free, three-week summer program for third through fifth graders in underserved communities

[21] IEEE Computer Society. (n.d.-a). *About IEEE Computer Society*. Retrieved October 20, 2020, from https://www.computer.org/about.

[22] IEEE Computer Society. (n.d.). *Women in Computing*. Retrieved October 20, 2020, from https://www.computer.org/communities/women-in-computing.

[23] National Society of Black Engineers. (n.d.). *About Us*. Retrieved October 20, 2020, from https://www.nsbe.org/About-Us/NSBE-Vision-Mission-Objectives.aspx#.X54jXNBKg2w.

[24] National Society for Black Engineers. (n.d.). *Women in Science and Engineering*. Retrieved October 20, 2020, from https://www.nsbe.org/Professionals/Programs/Special-Interest-Groups-(SIGs)/Women-in-Science-Engineering-(WiSE).aspx#.X54iddBKg2w.

across the country. While most of the curricula focus on engineering, SEEK students also learn how to write code and program robots, enhancing their awareness and knowledge of tech. Since 2007, SEEK has served over 25,000 students in nearly 30 cities. While most of the programs are co-ed, NSBE has also hosted all-girls programs in Washington, DC; Atlanta, Georgia; and Jackson, Mississippi. SEEK students have demonstrated increased proficiency in math, science, and engineering knowledge.

NSBE's Pre-College Initiative features more than 225 NSBE Jr. chapters that provide year-round learning to inspire and equip third through twelfth graders to pursue education and careers in STEM. Featuring curricula that include robotics, mathematics, and programming, NSBE Jr. students prepare for regional and national competitions and cultivate their interest in STEM in a supportive environment. As with SEEK, while most chapters are co-ed, several include girls only and are led by professional women of color.

Society for Advancement of Chicanos and Native Americans in Science

The Society for Advancement of Chicanos and Native Americans in Science (SACNAS) is an organization dedicated to achieving "True Diversity in STEM." SACNAS serves a growing community of over 20,000 supporters and more than 8,200 members, and has more than 115 student and professional chapters throughout the United States and Puerto Rico. SACNAS influences the STEM diversity movement through outreach and advocacy, promotion of STEM leaders, and the SACNAS National Diversity in STEM Conference. SACNAS is an inclusive organization dedicated to fostering the success of Chicanos/Hispanics and Native Americans, from college students to professionals, in attaining advanced degrees, careers, and positions of leadership in STEM.[25]

Society of Asian Scientists and Engineers

The Society of Asian Scientists and Engineers (SASE) was founded in 2007 and is dedicated to the advancement of scientists and engineers of Asian heritage in education and employment so that they can achieve their full career potential. In addition to professional development, SASE encourages members to contribute to the enhancement of the communities in which they live. SASE's mission is to prepare scientists and engineers of Asian heritage for success in the global

---

[25] Society for the Advancement of Chicano and Native Americans in Science. (n.d.). https://www.sacnas.org.

business world, celebrate diversity on campuses and in the workplace, and provide opportunities for members to make contributions to their local communities.[26]

Society of Hispanic Professional Engineers

The Society of Hispanic Professional Engineers (SHPE) is the nation's largest association dedicated to fostering Hispanic leadership in STEM fields. SHPE supports members through high school, college, graduate programs, and throughout their professional careers. High school students are introduced to the world of STEM at SHPE events and through mentorship and scholarship opportunities. College students can get involved with their schools' chapters—or start their own. Young professional members receive opportunities to grow, and seasoned innovators in the field mentor and give back.[27]

The SHPEtinas program accelerates and affirms Latinx women's representation at all levels of STEM corporate and academic leadership. Encouraging Latinx women to pursue higher education and careers in STEM recognizes the unique perspective they bring to solving the world's most pressing problems while creating new, influential role models for future leaders. The SHPEtina conference covers over 20 topics related to empowering Latinx women in STEM, including career development, communication, skill building, and leadership.[28]

Society of STEM Women of Color

The Society of STEM Women of Color (SSWOC) is a membership organization of STEM women of color from throughout the United States. Its members come from diverse racial and ethnic groups and are employed in a wide range of positions at virtually all of the major institutional types among U.S. colleges and universities.[29] SSWOC's mission is to achieve professional liberation, the freedom to pursue one's professional goals to the extent of one's own aptitude without externally imposed limitations, for all women, particularly those from historically underrepresented populations. SSWOC members use cultural, structural, and disciplinary sources of authority to produce new knowledge.[30] SSWOC

[26] Society of Asian Scientists and Engineers. (n.d.). *About SASE*. Retrieved October 31, 2020 from https://saseconnect.org/about-sase.

[27] Society of Hispanic Professional Engineers. (n.d.-a). *About*. Retrieved October 20, 2020 from https://www.shpe.org/about-shpe.

[28] Society of Hispanic Professional Engineers. (n.d.). *SHPEtinas*. Retrieved October 20, 2020 from https://www.shpe.org/shpetinas.

[29] Society of STEM Women of Color. (n.d.). *About*. Retrieved October 30, 2020 from https://www.sswoc.org/about-2/.

[30] Society of STEM Women Color. (n.d.). *Our Mission*. Society of STEM Women of Color. Retrieved October 30, 2020 from https://www.sswoc.org/about-2/our-mission/.

is home to a diverse group of members committed to an intersectional approach for empowering women of color in STEM.[31]

Society of Women Engineers

The mission of the Society of Women Engineers (SWE) is to empower women to achieve full potential in careers as engineers and leaders, expand the image of the engineering and technology professions as a positive force in improving the quality of life, and demonstrate the value of diversity and inclusion. The society is a non-profit, educational, and service organization that empowers women to succeed and advance in the field of engineering and be recognized for their life-changing contributions as engineers and leaders. SWE is the driving force establishing engineering as a highly desirable career for women through an array of training and development programs, networking opportunities, scholarships, and outreach and advocacy activities.[32]

RECOMMENDATIONS

In response to the insufficient number of workforce-ready individuals who can enter computing and technical occupations to meet the demand, there is a growth in industries that offer competency-based training for entry into their organizations, in badging and certification opportunities, and in community-based organizations offering these alternative pathways. These efforts can provide women of color, an untapped resource, with alternative entry points into tech occupations.

The recommendations that follow address the roles of academia, community organizations, industry, federal agencies, and professional societies in changing these numbers through education of K-12 students and retraining programs for adults. The recommendations are informed by the discussions and findings from the committee's workshops, the research literature, and data gathered over the course of the study.

RECOMMENDATION 6-1. Industry and funding agencies should invest in expansion of certification and training programs for women of color that are delivered by community-based organizations to scale their capacity to recruit and prepare a greater number of women of color in tech. These investments should expand opportunities for apprenticeships and people seeking to (re)enter the tech workforce.

---

[31] Society of STEM Women of Color. (n.d.-b). *Home Page*. Retrieved October 30, 2020, from https://www.sswoc.org/.

[32] Society of Women Engineers. (n.d.). *About SWE*. Retrieved October 10, 2020, from https://swe.org/about-swe/.

The low number of women of color in tech positions who have not received a bachelor's degree (Table 6-1) and who earn certificates (Figures 6-1 and 6-2) demonstrates that women of color are not taking sufficient advantage of alternative pathways into tech careers. Recently, there has been significant interest in reskilling in computing-related areas among non-computer science majors (NAS, NAE, and IOM, 2005; NASEM, 2018). Re-entry programs provide a substantial opportunity for women who stepped away from the workplace for family reasons and seek reentry, perhaps in a career they previously may not have considered.

Professional preparation of women of color can be an integral component of an organization's diversity, equity, and inclusion strategy. Dedicated efforts in areas of national need, such as artificial intelligence, cybersecurity, and data analytics, can provide entry into tech fields and provide women of color the appropriate knowledge, skills, and abilities that can lead to progressively more advanced roles.

RECOMMENDATION 6-2. Funding agencies should invest in programs that provide scholarships to Native female students who pursue a graduate program in a computing-related field and commit to teach at a tribal college or university for the length of the scholarship.

Tribal colleges and universities are an important entry point into technology and computing fields for Native female students; however, there are not many tribal colleges and universities that offer a bachelor's or master's degree; most are similar to a community college (Varma, 2009a, 2009b). While these institutions provide curricula aligned to the culture of American Indians and Alaska Natives (Ambler, 2002), institutions are influenced by the structural and geographical challenges experienced on reservations where they are located—for example, high unemployment rates, low per-capita income, lack of qualified instructors, and hard-to-reach locations (Varma, 2009a, 2009b).These factors represent barriers for prospective faculty to teach technology at these institutions. Only 12 out of 35 tribal colleges and universities offer career pathways in computing.

Providing incentives to acquire the credentials needed to teach at a tribal college or university could leverage a common desire of women of color to give back to their community. This desire connects to Carlone and Johnson's (2007) concept of the altruist scientist, whose scientific identity is tied to altruistic values connected to science as the means to improve people's lives. The literature demonstrates that many women of color consider altruistic values as an intrinsic part of their identity as scientists and seek to give back by supporting their communities, mentoring or serving as role models, and supporting those who are like them in some way, such as sharing their same gender and/or race/ethnicity or being interested in similar fields (Agbenyega, 2018; Foster, 2016; Herling, 2011;

Hodari et al., 2014, 2015, 2016; Lyon, 2013; Rodriguez, 2015; Skervin, 2015; Thomas, 2016).

The NSF Cybercorps® Scholarship for Service[33] program provides a model for increasing the number of computing programs offered at tribal colleges and universities. Scholarship for Service offers scholarships to students who pursue a post-baccalaureate degree in cybersecurity who commit to working for the federal, state, local, tribal, or territorial government, or a federally funded research and development center, after graduation for a period equal to the duration of the scholarship. Such a program could provide financial support for Native female students to seek a post-baccalaureate degree and give back to their community by becoming an instructor at a tribal college or university in a computing-related field.

> **RECOMMENDATION 6-3. Higher education administrators should incentivize technology and computing-related departments to accept tech-related certification and digital badges, and should provide well-defined pathways for women of color and others from technology training programs offered by community colleges, industry, and especially community-based organizations toward earning associates, undergraduate, and graduate degrees in tech fields.**

Industry-based training programs represent new pathways for employees and others to earn advanced degrees in technology. The programs provide articulated educational benefits through partnerships with higher education institutions such as Northeastern University, one of the first institutions to offer workplace badges for academic credit. It is not clear if women of color are taking advantage of these emerging pathways, perhaps because of the high barrier to entry into industry positions. On the other hand, community-based technology training programs, particularly those that target women of color, provide supportive environments for women to gain information technology skills and earn certifications and badges. These programs tailor their recruitment messaging, instruction, and wrap-around services and provide ongoing support for their alumna.

> **RECOMMENDATION 6-4. Professional societies should create programs and/or initiatives directed at developing additional pathways that advance women of color in tech. These programs should have a strong evaluation component to demonstrate impact and provide recommendations for scaling successful models. Programming should include certification and badging options defined collaboratively with, and recognized by, industry and academic partners.**

---

[33] For more information see https://beta.nsf.gov/funding/opportunities/cybercorps-scholarship-service-sfs-0.

Moreover, professional societies should be intentional about diversifying their internal leadership.

Professional societies support the development of standards and are positioned to "design and promote change, including through publications, policy statements, meetings, committees, lectureships, and awards" (NAS, NAE, and IOM, 2005). Furthermore, these societies often offer educational and informational resources and can offer support to students who are interested in educational and career opportunities in a specific discipline (Morris and Washington, 2017; NAS, NAE, and IOM, 2005). Unfortunately, this influence is not often exercised as effectively as it could or should be concerning increasing diversity. While professional societies may episodically focus their outreach to women and individuals from underrepresented groups, they often experience little success in increasing engagement or participation.

The real key to engaging and broadening participation is designing programs and initiatives that are shaped by and for the groups they purport to target (Morris and Washington, 2017). A review of the longstanding professional societies that support individuals in one or more of the STEM fields revealed no programs or initiatives focused specifically on creating pathways or advancing women of color in tech, though some have programs or initiatives for women in tech and/or STEM, and/or people of color in tech and/or STEM. Moreover, the leadership of professional societies rarely reflects the diversity of the future workforce. Even among professional societies that specifically serve people of color, there are very few with programming or initiatives solely for women of color in tech.

REFERENCES

Abbate, J., 2018. Code switch: Alternative visions of computer expertise as empowerment from the 1960s to the 2010s. *Technology and Culture* 59(4):S134-S159.

Abramovich, S., C. Schunn, and R. M. Higashi. 2013. Are badges useful in education? It depends upon the type of badge and expertise of learner. *Educational Technology Research and Development* 61(2):217-232.

ACTE (Association for Career and Technical Education). 2017. *Connecting industry-recognized certification data to education and workforce outcomes: Measuring the value added to skills, employment and wages.* Alexandria, VA.

Agbenyega, E. T. B. 2018. "We are fighters": Exploring how Latinas use various forms of capital as they strive for success in STEM. PhD dissertation, Temple University.

Ambler, M. 2002. Sustaining our home, determining our destiny. *Tribal College* 13(3):8. https://tribalcollegejournal.org/sustaining-home-determining-destiny/.

Beer, A. 2018. Apprenticeships: An emerging community college strategy for workforce development. Washington, DC: Association of Community College Trustees.

Besser, E. D., and T. J. Newby. 2020. Feedback in a digital badge learning experience: Considering the instructor's perspective. *TechTrends* 64(3):484-497.

Bureau of Labor Statistics, U.S. Department of Labor, Occupational Outlook Handbook, Computer and Information Research Scientists. https://www.bls.gov/ooh/computer-and-information-technology/computer-and-information-research-scientists.htm (visited October 2, 2021).

Carlone, H. B., and A. Johnson. 2007. Understanding the science experiences of successful women of color: Science identity as an analytic lens. *Journal of Research in Science Teaching* 44(8):1187-1218.

Carnevale, A. P., S. J. Rose, and A. R. Hanson. 2012. *Certificates: Gateway to gainful employment and college degrees.* Washington, DC: Georgetown University Center on Education and the Workforce.

Chapple, K. 2006. Low-income women, short-term job training programs, and IT careers. In *Women and information technology: Research on underrepresentation*, edited by J. M. Cohoon and W. Aspray. Cambridge, MA: MIT Press. Pp. 439-470.

Claussen, D. S. 2017. Digital badges in education: Trends, issues, and cases. *Journalism and Mass Communication Educator* 72(3):368-369.

Devedžić, V., and J. Jovanović. 2015. Developing open badges: A comprehensive approach. *Educational Technology Research and Development* 63(4):603-620.

Facey-Shaw, L., M. Specht, P. van Rosmalen, and J. Bartley-Bryan. 2020. Do badges affect intrinsic motivation in introductory programming students? *Simulation and Gaming* 51(1):33-54.

Foster, C. 2016. Hybrid spaces for traditional culture and engineering: A narrative exploration of Native American women as agents of change. PhD dissertation. Arizona State University, Tempe, Arizona.

Gibson, D., N. Ostashewski, K. Flintoff, S. Grant, and E. Knight. 2015. Digital badges in education. *Education and Information Technologies* 20(2):403-410.

Ham, B. 2020. AAAS drafts plan to address systemic racism in sciences. https://www.science.org/doi/pdf/10.1126/science.370.6516.541.

Herling, L. 2011. *Hispanic women overcoming deterrents to computer science: A phenomenological study.* Vermillion, SD: University of South Dakota.

Higashi, R., and C. D. Schunn. 2020. Perceived relevance of digital badges predicts longitudinal change in program engagement. *Journal of Educational Psychology* 112(5):1020.

Hodari, A. K., M. Ong, L. T. Ko, and R. R. Kachchaf. 2014. New enactments of mentoring and activism: U.S. women of color in computing education and careers. Proceedings of the 10th Annual Conference on International Computing Education Research. Pp. 83-90.

Hodari, A. K., M. Ong, L. T. Ko, and J. Smith. 2015. Enabling courage: Agentic strategies of women of color in computing. In *2015 Research in Equity and Sustained Participation in Engineering, Computing, and Technology (RESPECT)*, Institute of Electrical and Electronics Engineers. Pp. 1-7. doi: 10.1109/RESPECT.2015.7296497.

Hodari, A. K., M. Ong, L. T. Ko, and J. M. Smith. 2016. Enacting agency: The strategies of women of color in computing. *Computing in Science and Engineering* 18(3):58-68.

Hu, J. C. 2019. NSF graduate fellowships disproportionately go to students at a few top schools. https://www.science.org/content/article/nsf-graduate-fellowships-disproportionately-go-students-few-top-schools.

Kappes, S., and V. C. Betro. 2015. Using Mozilla badges to certify XSEDE users and promote training. Proceedings of the 2015 XSEDE Conference: Scientific Advancements Enabled by Enhanced Cyberinfrastructure. Pp. 1-4.

Kvasny, L. 2006. Let the sisters speak: Understanding information technology from the standpoint of the "other." *ACM SIGMIS Database* 37(4):13-25.

Kvasny, L., and J. Chong. 2006. Third World feminist perspectives on information technology. In *Encyclopedia of gender and information technology* (pp. 1166-1171). IGI Global.

Leaser, D., K. Jona, and S. Gallagher. 2020. Connecting workplace learning and academic credentials via digital badges. *New Directions for Community Colleges* 189:39-51.

Lerman, R. I. 2016. Restoring opportunity by expanding apprenticeship. In *The dynamics of opportunity in America*, edited by I. Kirsch and H. Braun. Springer. Pp. 359-385.

Lipman, J. 2019. Helping stay-at-home parents reenter the workforce. *Harvard Business Review.* June 7. https://hbr.org/2019/06/helping-stay-at-home-parents-reenter-the-workforce.

Lyon, L. A. 2013. Sociocultural influences on undergraduate women's entry into a computer science major. PhD dissertation. University of Washington, Seattle, WA.

Mardis, M. A., J. Ma, F. R. Jones, C. R. Ambavarapu, H. M. Kelleher, L. I. Spears, and C. R. McClure. 2018. Assessing alignment between information technology educational opportunities, professional requirements, and industry demands. *Education and Information Technologies* 23(4):1547-1584.

Moore, A. 2018. Do you have a badge for that? Association for Talent Development, February 1. https://www.td.org/magazines/td-magazine/do-you-have-a-badge-for-that.

Morris, V. R., and T. M. Washington. 2017. The role of professional societies in STEM diversity. *Journal of the National Technical Association* 87(1):22-31.

NAS, NAE, and IOM (National Academy of Sciences, National Academy of Engineering, and Institute of Medicine). 2005. *Facilitating interdisciplinary research.* Washington, DC: The National Academies Press. https://doi.org/10.17226/11153.

NASEM (National Academies of Sciences, Engineering, and Medicine). 2018. *Assessing and responding to the growth of computer science undergraduate enrollments.* Washington, DC: The National Academies Press. https://doi.org/10.17226/24926.

NCES (National Center for Education Statistics). IPEDS: Integrated Postsecondary Education Data System (2014-2015 to 2018-2019, Certificates awarded to women of color based on type of certificate; accessed August 20, 2020). https://nces.ed.gov/ipeds/.

Newby, T. J., and Z. Cheng. 2020. Instructional digital badges: Effective learning tools. *Educational Technology Research and Development* 68(3):1053-1067.

Pitt, C. R., A. Bell, R. Strickman, and K. Davis. 2019. Supporting learners' STEM-oriented career pathways with digital badges. *Information and Learning Sciences* 120(1/2):87-107. https://doi.org/10.1108/ILS-06-2018-0050.

Rehak, A. M., and D. T. Hickey. 2013. A multi-level analysis of engagement and achievement: Badges and wikifolios in an online course. Presentation at the Computer-Supported Collaborative Learning Conference, Madison, WI, June 15-16, 2013.

Rodriguez, S. 2015. Las mujeres in the STEM pipeline: How Latina college students who persist in STEM majors develop and sustain their science identities. PhD dissertation. University of Texas, Austin.

Shah, S. 2020. *Breaking through, rising up: Strategies for propelling women of color in technology.* Brooklyn, NY: NPower.

Skervin, A. 2015. *Success factors for women of color information technology leaders in corporate America.* Minneapolis, MN: Walden University.

Thomas, S. S. 2016. An examination of the factors that influence African American females to pursue postsecondary and secondary information communications technology education. PhD dissertation. Texas A&M University, College Station. http://hdl.handle.net/1969.1/156994.

Tribal College Journal. 2019. Degree Programs at Tribal Colleges and Universities 2019-2020. Mancos, CO: Tribal College Journal. https://tribalcollegejournal.org/pdfs/Degree-Programs-at-Tribal-Colleges-and-Universities.pdf.

Van Noy, M., M. Trimble, D. Jenkins, E. Barnett, and J. Wachen. 2016. Guided pathways to careers: Four dimensions of structure in community college career-technical programs. *Community College Review* 44(4):263-285.

Varma, R. 2009a. Attracting Native Americans to computing. *Communications of the ACM* 52(8):137-140.

Varma, R. 2009b. Bridging the digital divide: Computing in tribal colleges and universities. *Journal of Women and Minorities in Science and Engineering* 15(1):39-52.

Voogt, L., L. Dow, and S. Dobson. 2016. Open badges: A best-practice framework. *Proceedings of the 2016 SAI Computing Conference.* Institute of Electrical and Electronics Engineers. Pp. 796-804.

Wachen, J., D. Jenkins, and M. Van Noy. 2010. How I-BEST Works: Findings from a Field Study of Washington State's Integrated Basic Education and Skills Training Program, Community College Research Center, Columbia University.

Wardrip, P. S., S. Abramovich, Y. J. Kim, and M. Bathgate. 2016. Taking badges to school: A school-based badge system and its impact on participating teachers. *Computers and Education* 95:239-253.

Zeidenberg, M., S. W. Cho, and D. Jenkins. 2010. *Washington State's Integrated Basic Education and Skills Training Program (I-BEST): New Evidence of Effectiveness.* CCRC Working Paper No. 20. Community College Research Center, Columbia University.

Zhang, J., and C. Oymak. 2018. Participants in sub-baccalaureate occupational education: 2012. Stats in Brief. NCES 2018-149. Washington, DC: US Department of Education. https://nces.ed.gov/pubs2018/2018149.pdf.

Appendix A

Alliances Focused on Women of Color and Underrepresentation in Tech

The appendices list alliances, professional organizations, and programs that focus on women of color and underrepresentation in tech. The alliances listed here are focused on broadening participation in computing.

CENTER FOR MINORITIES AND PEOPLE WITH DISABILITIES IN IT (CMD-IT)

CMD-IT is comprised of corporations, academic institutions, government agencies, and non-profits with a mission to ensure that underrepresented groups are fully engaged in computing and information technologies, and to promote innovation that enriches, enhances, and enables these communities, such that more equitable and sustainable contributions are possible by all communities. Programs include professional development workshops, mentorships workshops, scholarship, and an annual Tapia Conference (created to bring together undergraduate and graduate students, faculty, researchers, and professionals in computing from all backgrounds and ethnicities).

COMPUTING ALLIANCE OF HISPANIC-SERVING INSTITUTIONS (CAHSI)

CAHSI, an NSF national INCLUDES alliance, is a national network of academic institutions, non-profits, industry leaders, and governmental entities working to advance Hispanics in computing in the workforce and academia. The mission of CAHSI is to grow and sustain a networked community committed to recruiting, retaining, and accelerating the progress of Hispanics in computing.

CAHSI's programming focused on the advancement of Latinas includes CAHSI Student Advocates—undergraduate students who connect students with CAHSI opportunities and CAHSI leadership; CAHSI Latina Scholars—undergraduate or graduate Latinas who are recognized for their outstanding achievements in the classroom and in the community and who will be the next generation of Latina leaders in technology; and the Peer Allyship program that pairs lower division Latinas with upper division Latina students with the aim of building their networks and success strategies. Other signature practices that impact Latinas is FemProf, a program that prepares junior female students for graduate studies and entry into the professoriate, and the Affinity Research Group model that focuses on the deliberate development of research, professional, and communication skills.

ECEP: EXPANDING COMPUTING EDUCATION PATHWAYS

The ECEP Alliance seeks to increase the number and diversity of students in the pipeline to computing and computing-intensive degrees by supporting state-level computing education reforms. Through interventions, pathways, partnerships and models that drive state-level computing education change, ECEP supports states as they work to align their state efforts with the national vision for computer science for all.

INSTITUTE FOR AFRICAN-AMERICAN MENTORING IN COMPUTER SCIENCES

iAAMCS serves as a national resource for all African American computer science students and faculty. It aims to increase the number of African Americans receiving Ph.D. degrees in computing sciences, promote and engage students in teaching and training opportunities, and add more diverse researchers into the advanced technology workforce.

NATIONAL CENTER FOR WOMEN & INFORMATION TECHNOLOGY (NCWIT)

NCWIT is a non-profit community that convenes, equips, and unites change leader organizations to increase the influential and meaningful participation of girls and women—at the intersections of race, ethnicity, class, age, sexual orientation, and disability status—in the field of computing, particularly in terms of innovation and development. NCWIT uses a three-pronged strategy: (1) NCWIT brings together change leaders who carry out projects and initiatives in support of NCWIT's mission from universities, companies, non-profits, and government organizations; (2) NCWIT provides free, online research-based resources for reform at every level to help individuals implement change, raise awareness, and reach out to critical populations through outreach events and members' networks;

and (3) NCWIT develops programs for members to achieve goals focused on policy reform, image change, outreach to underrepresented groups, etc. Such programs include NCWIT Aspirations in Computing (an initiative that provides technical girls and women with ongoing engagement, visibility, and encouragement for their computing-related interests and achievements from high school through college and into the workforce) and Sit With Me (a fun, creative national advocacy campaign that uses an iconic red chair to symbolize the critical need for women's technical contributions).

STARS COMPUTING CORPS

The STARS Computing Corps is a national alliance with a mission to broaden participation of underrepresented groups in computing within institutions of higher education. In particular, STARS aims to increase computing persistence and promote career advancement for undergraduates, graduate students, and faculty, with a focus on addressing systemic and social barriers faced by those from underrepresented groups in computing.

Appendix B

Professional Organizations and Programs Focused on Women of Color and Underrepresentation in Tech

ANITAB.ORG

AnitaB.org supports women in technical fields, as well as the organizations that employ them and the academic institutions training the next generation. A full roster of programs helps women to grow, learn, and develop their highest potential. AnitaB.org provides women in tech with year-round opportunities to connect with and inspire one another, develop their professional skills, find mentors, and gain recognition. The organization's communities, events, and programs offer the resources women need to build rewarding careers in technology. AnitaB. org is driven by the belief that more can be accomplished together. To address the holistic needs of women and non-binary technologists, the organization created a membership program with resources, opportunities, and connections to support every career level. AnitaB.org works with organizations and individuals to identify and overcome industry challenges, diversify workforces, and foster cultures where women technologists create impactful and lasting careers.

ASSOCIATION FOR WOMEN IN SCIENCE (AWIS)

"AWIS champions the interests of women in science, technology engineering, and mathematics across all disciplines and employment sectors. Working for positive system transformation, AWIS strives to ensure that all women in these fields can achieve their full potential."

ASSOCIATION FOR WOMEN IN MATHEMATICS

"The purpose of the Association for Women in Mathematics is to encourage women and girls to study and to have active careers in the mathematical sciences, and to promote equal opportunity and the equal treatment of women and girls in the mathematical sciences."

BLACK COMPUTEHER

Black ComputeHer is dedicated to supporting computing+tech education and workforce development for Black women and girls. The organization aims to create rich technical programming, lead empirical research, and disseminate information that addresses the lack of inclusive innovation in tech across education and industry. There is an abundance of literature that examines why women do not pursue computer science (CS), and research that investigates why Black students do not pursue CS. Unfortunately, according to Black ComputerHer, there is not much research that addresses the impact of the intersectionality of gender, race, and other constructs (socioeconomics, regional experiences, educational experiences, etc.) on Black women along the pathway of success in computing. As such the organization is conducting research that contributes to this growing body of knowledge.

BLACK FEMALE FOUNDERS

The mission of Black Female Founders is to provide awareness, promotion, support, and resources for Black women-led tech-based and tech-enabled startups throughout the U.S. and Black Diaspora. The organization's platform and programs advance women from aspiration to investment. In an effort to increase diversity and inclusion within the global innovation ecosystem, Black Female Founders elevate and empower tech-enabled or tech-based ventures from idea conception to business implementation and toward investment. The organization's platform is designed to increase exposure of Black female founders and their ventures. Founders receive help with business creation and development via programs, access to investors, and more. The organization's signature program, the #BFF Labs pre-accelerator, provides critical industry knowledge, mentorship, and business development tools to help fledgling startups succeed. #BFF was created by Black women who are also entrepreneurs with experience in the tech, finance, and investment space.

BLACK WOMEN IN SCIENCE AND ENGINEERING (BWISE)

BWISE was founded in 2015 with the purpose to support underrepresented women in bridging the leadership gap through networking, mentorship, and career development. The group consists of Black women from middle management through senior leadership with degrees in the sciences, math, and engineering. The BWISE mission is to empower Black women through career and entrepreneurial development, insight, and training. The BWISE vision is to significantly impact the diversity of the STEM pipeline, both corporate and academic, from beginning to end.

BLACK WOMEN IN TECHNOLOGY

Black Women in Technology was founded in 2014 in Los Angeles, California and was created to engage Black women to enter tech as a career choice. Black Women in Technology's mission is to teach tech invention in order to serve our community, engage Black women and women of color to embrace tech innovation and serve one another through positive role models and relationships.

BLACKS IN TECHNOLOGY LLC (BIT)

BIT is the largest community and media organization that focuses on Black people in the technology industry. Through community-focused activities, events and media, Blacks In Technology (BIT) is "Stomping the Divide" by establishing a blueprint of world-class technical excellence and innovation by providing resources, guidance and issuing a challenge to our members to surpass the high mark and establish new standards of global innovation.

BLACK GIRLS CODE

The vision of Black Girls CODE is to increase the number of women of color in the digital space by empowering girls of color ages 7 to 17 to become innovators in STEM fields, leaders in their communities, and builders of their own futures through exposure to computer science and technology. The organization works to provide African American youth with the skills to occupy some of the 1.4 million computing job openings expected to be available in the United States by 2020, and to train 1 million girls by 2040. Black Girls CODE introduces computer coding lessons to young girls from underrepresented communities in programming languages such as Scratch or Ruby on Rails.

COMPUTER RESEARCH ASSOCIATION (CRA)

CRA's Committee on Widening Participation in Computing Research (CRA-WP) has a mission to increase the success and participation of women, underrepresented minorities, and persons with disabilities in computing research and education at all levels. CRA-WP programs, people, and materials provide mentoring and support for women, underrepresented minorities, and persons with disabilities at every level of the research pipeline: undergraduate students, graduate students, faculty, and government and industry researchers.

G|CODE

G|Code is a Boston-based non-profit organization that offers young women of color a safe co-living, working, and learning community where they will learn cutting-edge technology skills, gain employment experience, and connect with our world renowned network of mentors, advisors, and enterprise partners. Intro to G|Code is a 10-week program for young women of color, created by young women of color. The program provides opportunity for 16 young women (known as the ChangeMakers) to learn about tech, foster community, practice self-reflection, and prepare for their futures. This program exposes ChangeMakers to software and development fundamentals, as well as personal and professional development that is key to successful careers and general well-being. Following completion of Intro to G|Code, ChangeMakers may go on to participate in further development with allied programs, i.e., Resilient Coders, Year Up, Hack Diversity, Apprenti, and others. G|Code House is 24-month program includes nine months of in-class technical training, six months of an internship/co-op, nine months of specialty training (e.g., cyber security), and mentorship and personal support in a safe, focused home.

GIRLS IN TECH

Founded in 2007 by Adriana Gascoigne, Girls in Tech is a non-profit organization dedicated to eliminating the gender gap in tech. They have more than 60,000 members in 50+ chapters around the world. Girls in Tech started with an idea: There is a strong, smart, and outspoken girl within all of us. Girls in Tech exists to make sure that girl is heard. Because when every voice, every perspective, every personality is honored and respected, we do better work and live richer lives. Girls in Tech is passionate about inclusivity, because they know that tech today requires people of all skills and backgrounds. They are committed to building the diverse and inclusive tech workforce the world needs. And they aim to see every person accepted, confident, and valued in tech—just as they are.

GREAT MINDS IN STEM (GMIS)

GMiS is the gateway for Hispanics in science, technology, engineering and mathematics (STEM). Established in 1989, as HENAAC, Great Minds in STEM is a non-profit organization that focuses on STEM educational awareness programs for students from kindergarten to career. Great Minds in STEM provides resources for recognition and recruitment of Hispanics in STEM on a national level, connecting multi-areas of engineering and science arenas to the general population.

LATINAS IN COMPUTING

Latinas in Computing (LiC) is a community created by and for Latinas with a mission of promoting Latina representation and success in computing-related fields. The LiC community was established with the help of AnitaB.org after a Birds of a Feather session at the 2006 Grace Hopper Celebration of Women in Computing. LiC is part of AnitaB.org Systers Affinity Groups and was established to inspire the community of Latinas in computing, promote opportunities and resources, provide resources and guidance to lead new initiatives, and mentor students, junior faculty, and professionals.

LATINAS IN TECH

Latinas in Tech is a non-profit organization with the aim to connect, support, and empower Latina women working in tech, including (1) to increase the number of Latinas working in the tech industry; (2) to increase representation of Latinas in tech in decision-making positions; (3) to increase the participation of Latina-founded startup in Venture Capital funding of $1M+$; and (4) to increase Latinas' confidence in their own capabilities and skills. They work with top technology companies to create safe spaces for learning, mentorship, and recruitment. They provide networking meetups; webinars around career development and skill building; mentorship program and training; leadership trainings for entry, mid, and executive level Latinas; recruiting events; and an annual summit.

NATIVE AMERICAN WOMEN IN COMPUTING (NAWiC)

NAWiC is a community that brings support and inspiration to Indigenous women in technical fields across North and South America. Native women need a safe space to share their work, ideas, experiences, and accomplishments and get recognition for them. There are Native women currently in the tech industry, learning to code. NAWiC is a community to bridge that gap between a new coder and a tech executive, and everything in between. NAWiC is working on reaching out to Native/Indigenous communities to host workshops, meetups, and hackathons.

REBOOT REPRESENTATION TECH COALITION

A coalition of tech companies that have come together to align funding and agendas to address the lack of representation, particularly of women of color, within the technology industry. Pivotal Ventures catalyzed the formation of the coalition through the development of the Rebooting Representation report (co-authored by McKinsey & Co.) and the coalition is housed at the National Center for Women & Information Technology. The goal is to double the number of

Black, Latinx, and Native American women receiving computing degrees by 2025. The coalition does this by (1) pooling corporate dollars in order to make strategic investments in organizations that provide rigorous computer science education for Black, Latinx, and Native American women; (2) raising the profile of the issue of underrepresentation of women of color in computing through frequent communications activities; and (3) bringing companies together to share best practices and lessons learned and to use one unified voice.

WOMEN IN TECHNOLOGY

Women in Technology (WIT) has the sole aim of advancing women in technology—from the classroom to the boardroom. WIT meets its vision through a variety of leadership development, technology education, networking, and mentoring opportunities for women at all levels of their careers. WIT has over 1000 members in the Washington, D.C./Maryland/Virginia metro region.

Appendix C

Workshop Agendas

**COMMITTEE ON ADDRESSING THE UNDERREPRESENTATION
OF WOMEN OF COLOR IN TECH**

**PUBLIC WORKSHOP #1
FEBRUARY 5, 2020**

**Keck Center of the National Academies of
Sciences, Engineering, and Medicine
Room 100**
500 Fifth Street, NW
Washington, DC 20001

9:00 – 9:10 am **Welcome and Opening Remarks**
Valerie Taylor and Evelynn Hammonds

9:10 – 9:40 am **Morning Keynote Address**
• ***Kyla McMullen****, Associate Professor and Director of
SoundPad Lab, University of Florida*

9:40 – 10:10 am **Discussion and Q&A**
Moderator: Valerie Taylor, Committee Co-Chair

10:10 – 10:20 am **BREAK**

10:20 – 11:05 am   PANEL 1: Underrepresentation of Women of Color
in Tech: Industry
Moderator: Manuel Pérez-Quiñones,
Committee Member
- *Jamika Burge, Senior Manager, Capital One*
- *Dora Renaud, Senior Director of Academic and
  Professional Development Programs, Society of
  Hispanic Professional Engineers*
- *Rati Thanawala, Advanced Leadership Fellow,
  Harvard University*

11:05 – 11:35 am   Panel Discussion and Q&A

11:35 – 12:30 pm   LUNCH

12:30 – 1:00 pm   Afternoon Keynote Address
- *Mia Ong, Senior Research Scientist and Evaluator, TERC*

1:00 – 1:30 pm   Discussion and Q&A
Moderator: Evelynn Hammonds, Committee Co-Chair

1:30 – 2:30 pm   PANEL 2: Underrepresentation of Women of Color in
Tech: Higher Education and Academia
Moderator: Allison Scott, Committee Member
- *Stephanie Adams, Dean and Lars Magnus Ericsson
  Chair, Eric Jonsson School of Engineering and
  Computer Science, University of Texas at Dallas*
- *Ann Gates, Chair and Professor, Computer Science
  Department, University of Texas at El Paso*
- *Joan Reede, Dean for Diversity and Community
  Partnership, Harvard University*
- *JeffriAnne Wilder, Senior Research Scientist, National
  Center for Women & Information Technology*

2:30 – 3:00 pm   Panel Discussion/Q&A

3:00 – 3:15 pm   BREAK

3:15 – 3:45 pm   PANEL 3: Underrepresentation of Women of Color in
Tech: The Role of Federal Agencies
Moderator: Frances Colón, Committee Member
- *Julie Carruthers, Senior Science and Technology
  Advisor, Office of the Deputy Director for Science
  Programs, DOE*

- *Evelyn W. Kent*, *Office of the Undersecretary of Defense for Research and Engineering*

3:45 – 4:15 pm   **Panel Discussion/Q&A**

4:15 – 4:25 pm   **Closing Remarks**
- *Fay Cobb Payton*, *Program Director, Computer and Information Science and Engineering Directorate, National Science Foundation*

4:25 – 4:30 pm   **Meeting Wrap-Up**
Valerie Taylor and Evelynn Hammonds

COMMITTEE ON ADDRESSING THE UNDERREPRESENTATION OF WOMEN OF COLOR IN TECH

PUBLIC WORKSHOP #2
Virtual Meeting

Day 1: April 7, 2020

12:00 – 12:15 pm **Welcome Remarks**, Evelynn Hammonds, Ph.D.

12:15 – 12:25 pm **Mary Schmidt Campbell, Ph.D.**, President of Spelman College

12:25 – 1:45 pm   **Keynote Address: Shirley Malcom, Ph.D.**, Head of Education and Human Resources Programs, American Association for the Advancement of Science
Q&A Moderated by Dr. Evelynn Hammonds, Harvard University

1:45 – 3:00 pm   **PANEL 1: Retention of Women of Color Students in Tech: An MSI Perspective**
- **Gloria Washington, Ph.D.**, Assistant Professor in Department of Computer Science, Howard University
- **Monique Ross, Ph.D.**, Assistant Professor, School of Computing and Information Sciences and STEM Transformation Institute, Florida International University

- **Raquel Hill, Ph.D.**, Associate Professor and Chair of Computer and Information Sciences, Spelman College

Q&A Moderated by Dr. Valerie Taylor, Argonne National Laboratory

3:00 pm   **Day 1 Adjourns**

Day 2: April 8, 2020

11:00 – 11:15 am   **Welcome Remarks, Dr. Valerie Taylor**

11:15 – 12:30 pm   **PANEL 2: Organizational Change and Women of Color in Tech**
- **Enobong "Anna" Branch, Ph.D.**, Vice Chancellor for Diversity, Inclusion, and Community Engagement and a Professor of Sociology, Rutgers University
- **Kaye Husbands Fealing, Ph.D.**, Chair and Professor, School of Public Policy, Georgia Institute of Technology

Q&A Moderated by Dr. Evelynn Hammonds, Harvard University

12:30 – 12:45 pm   **Closing Remarks, Fay Cobb Payton, Ph.D.**, Program Director, National Science Foundation

12:45 – 2:00 pm   **Day 2 Adjourns**

COMMITTEE ON ADDRESSING THE UNDERREPRESENTATION OF WOMEN OF COLOR IN TECH

PUBLIC WORKSHOP #3
Virtual Meeting

Day 1: May 14, 2020

12:00-12:15pm EDT   **WELCOME REMARKS**
Fay Cobb Payton, Program Director, Division of Computer and Network Systems, National Science Foundation

12:15-1:30pm EDT   **PANEL SESSION 1**
- **Timothy Pinkston**, Professor and George Pfleger Chair in Electrical and Computer Engineering, and Vice Dean for Faculty Affairs, University of Southern California
- **Ashley Carpenter**, Program Coordinator of Diversity Initiatives and University Center for Exemplary Mentoring (UCEM), Massachusetts Institute of Technology
- **Beena Sukumaran**, Vice President for Research, Rowan University
- **Kamau Bobb**, Senior Director, Constellations Center for Equity in Computing, Georgia Tech
Session Moderator: Gilda Barabino, Daniel and Frances Berg Professor and Dean of The Grove School of Engineering, City College of New York

1:30-2:45pm EDT   **PANEL SESSION 2**
- **Denise Peck**, Executive Advisor, Ascend Leadership
- **Dilma Da Silva**, Professor and Associate Dean, Texas A&M University; Computing Research Association
- **Cherri Pancake**, President, Association for Computing Machinery
- **Rocío Medina van Nierop**, Co-Founder and Executive Director, Latinas in Tech
Session Moderator: Ed Lazowska, Bill & Melinda Gates Chair in Computer Science & Engineering, University of Washington

2:45-3:00pm EDT   **CLOSING REMARKS**
Sarita E. Brown, President, Excelencia in Education

3:00pm EDT   **ADJOURN**

Day 2: May 15, 2020

12:00-12:15pm EDT   **WELCOME REMARKS**
Cynthia E. Winston-Proctor, Professor, Howard University Department of Psychology

12:15-1:30pm EDT   PANEL SESSION 3
- **Melonie Parker**, Chief Diversity Officer, Google
- **Bo Young Lee**, Chief Diversity and Inclusion Officer, Uber
- **Stephanie Lampkin**, Founder and Chief Executive Officer, Blendoor

Session Moderator: Karl W. Reid, Executive Director, National Society of Black Engineers

1:30-2:45pm EDT   PANEL SESSION 4
- **Renee Wittemyer**, Director of Program Strategy and Investment, Pivotal Ventures
- **Dwana Franklin-Davis**, Chief Executive Officer, Reboot Representation
- **Carlotta M. Arthur**, Director, Clare Boothe Luce Program for Women in STEM, Henry Luce Foundation
- **Carolina Huaranca Mendoza**, Founder, 1504 Ventures & Former Principal at Kapor Capital

Session Moderator: Manuel A. Pérez-Quiñones, Professor, Department of Software and Information Systems, University of North Carolina

2:45-3:00pm EST   CLOSING REMARKS
Evelynn M. Hammonds, Dean of Harvard College & Barbara Gutmann Rosenkrantz Professor of the History of Science and of African and African American Studies, Harvard University

3:00pm EST   ADJOURN

COMMITTEE ON ADDRESSING THE UNDERREPRESENTATION OF WOMEN OF COLOR IN TECH

PUBLIC WORKSHOP #4
Virtual Meeting

Day 1: June 3, 2020

12:00-12:15pm   WELCOME REMARKS
Sarah Echohawk, Chief Executive Officer, American Indian Science and Engineering Society

12:15-1:30pm **PANEL SESSION 1: Women of Color in Tech: A Focus on Native Americans in Computing**
- **Andrea Delgado-Olson**, Founder and Chair, Native American Women in Computing
- **Kathy DeerInWater**, Chief Program Officer, American Indian Science and Engineering Society
- **Twyla Baker**, President, Nueta Hidatsa Sahnish College
- **Sandra Begay**, Principal Member of the Technical Staff – Engineer, Sandia National Laboratories

Session Moderator: Sarah Echohawk, Chief Executive Officer, American Indian Science and Engineering Society

1:30-2:45pm **KEYNOTE PRESENTATION: Addressing the "Small-N" Problem**
- **Alice Pawley**, Associate Professor, School of Engineering Education, Purdue University

Session Moderator: Evelynn M. Hammonds, Barbara Gutmann Rosenkrantz Professor of the History of Science, Professor of African and African American Studies Harvard University

2:45-3:00pm **CLOSING REMARKS**

3:00pm **ADJOURN**

Day 2: June 4, 2020

12:00-12:15pm **WELCOME REMARKS**
Mia Ong, Senior Research Scientist and Evaluator, TERC

12:15-1:30pm **PANEL SESSION 2: Alternative Pathways to Careers in Tech for Women of Color**
- **Bridgette Wallace**, Founder and Executive Director, G|Code
- **Kenneth Walker**, Senior Vice President, Core Mission Support, Per Scholas
- **Jennifer Carlson**, Co-Founder, Executive Director, WTIA Workforce Institute and Apprenti

Session Moderator: Mia Ong, Senior Research Scientist and Evaluator, TERC

1:30-2:45pm   **PANEL SESSION 3: Supporting Women of Color in**
                       Tech Throughout Their Careers
                       • **Gregory M. Walton**, The Michael Forman University
                         Fellow in Undergraduate Education and Associate
                         Professor of Psychology, Stanford University
                       • **Marisela Martinez-Cola**, Assistant Professor of
                         Sociology, Utah State University
                       • **Nizhoni Chow-Garcia**, Associate Director, Office
                         of Inclusive Excellence, California State University,
                         Monterey Bay
                       <u>**Session Moderator:**</u> **Valerie Taylor**, Director,
                       Mathematics and Computer Science Division, Argonne
                       National Laboratory; CEO & President, Center for
                       Minorities and People with Disabilities in IT

2:45-3:00pm   **CLOSING REMARKS**
                       Fay Cobb Payton, Program Director, Division of
                       Computer and Network Systems, National Science
                       Foundation

3:00pm   **ADJOURN**

Appendix D

Committee Member Biographies

Evelynn M. Hammonds (Co-Chair) is the Barbara Gutmann Rosenkrantz Professor of the History of Science and Professor of African and African American Studies at Harvard University. She holds a B.S. in physics from Spelman College, a B.E.E. in electrical engineering from Georgia Tech, S.M. in physics from MIT and a Ph.D. in the history of science from Harvard University. Hammonds was the first Senior Vice Provost for Faculty Development and Diversity at Harvard University (2005-2008). From 2008 to 2013 she served as Dean of Harvard College. Her areas of research include the histories of science, medicine, and public health in the United States; race and gender in science studies; feminist theory; and African American history. Hammonds' work focuses on the intersection of genetic, medical, and sociopolitical concepts of race in the United States. She is currently the director of the Project on the Study of Race & Gender in Science & Medicine at the Hutchins Center for African and African American Research at Harvard University. Professor Hammonds is an Area Advisor for African American History, History of Science and Technology for the Online Bibliography of Oxford University Press.

Valerie Taylor (Co-Chair) is the Director of the Mathematics and Computer Science Division. She received her Ph.D. in electrical engineering and computer science from the University of California, Berkeley, in 1991. She then joined the faculty in the Electrical Engineering and Computer Science Department at Northwestern University, where she was a member of the faculty for 11 years. In 2003, she joined Texas A&M, where she served as head of the computer science and engineering department and senior associate dean of academic affairs in the College of Engineering and a Regents Professor and the Royce E. Wisenbaker

Professor in the Department of Computer Science. Her research is in the area of performance analysis and modeling of parallel, scientific applications. Currently, she is focused on the areas of performance analysis, power analysis, and resiliency. Dr. Taylor is the Executive Director of the Center for Minorities and People with Disabilities in IT (CMD-IT). The organization seeks to develop the participation of minorities and people with disabilities in the IT workforce in the United States.

Gilda A. Barabino is President of Olin College of Engineering. She previously served as Daniel and Frances Berg Professor and Dean at The City College of New York's (CCNY) Grove School of Engineering. Prior to joining CCNY, she was associate chair for graduate studies and professor in the Wallace H. Coulter Department of Biomedical Engineering at Georgia Tech and Emory. At Georgia Tech she also served as the inaugural Vice Provost for Academic Diversity. Prior to Georgia Tech and Emory, she rose to the rank of professor of chemical engineering and was Vice Provost for Undergraduate Education at Northeastern University. She is a noted investigator in the areas of sickle cell disease, cellular and tissue engineering, and the role of race/ethnicity and gender in science and engineering. Dr. Barabino is a member of the National Academy of Engineering and the National Academy of Medicine, a fellow of the American Association for the Advancement of Science (AAAS), the American Institute of Chemical Engineers (AIChE), the American Institute for Medical and Biological Engineering (AIMBE), and the Biomedical Engineering Society (BMES). Dr. Barabino serves on the Scientific Advisory Board of the Chan Zuckerberg Biohub. She is past-president of BMES and past-president of AIMBE. She has received an honorary degree from Xavier University of Louisiana, the Presidential Award for Excellence in Science, Mathematics and Engineering Mentoring, AIChE's Award for Service to Society, and AIMBE's Pierre Galetti Award. Dr. Barabino is a member of the National Science Foundation's Advisory Committee for Engineering, the congressionally mandated Committee on Equal Opportunities in Science and Engineering, the AAAS Committee on Science, Engineering and Public Policy, and the incoming chair of the National Academies Committee on Women in Science, Engineering, and Medicine. She consults nationally and internationally on STEM education and research, diversity in higher education, policy, faculty and workforce development. She received a B.S. from Xavier University of Louisiana and a Ph.D. from Rice University.

Sarita Brown is the co-founder and President of Excelencia in Education. For more than 30 years, she has worked at prominent educational institutions and at the highest levels of government to implement effective strategies to raise academic achievement and increase opportunity for low-income and students of color. She started her career at The University of Texas at Austin by building a national model promoting minority success in graduate education. Coming to the nation's capital to work for educational associations, Brown was tapped to serve as Executive Director of the White House Initiative for Educational Excellence for Hispanic Americans under President Bill Clinton and U.S. Secretary of Education Richard Riley. She later applied her talents and experience to the not-for-profit sector.

Jamika D. Burge is a senior manager at Capital One, where she oversees research curriculum development and internal/external outreach. At Capital One, she ideates and creates innovative user research curricula that empower designers, developers, and engineers to apply design thinking and human-centered design principles into their daily work, and beyond. She also engages with internal and external organizations in computing and design outreach activities. Prior to joining Capital One, Jamika served non-profit and government organizations, including as a consultant to the Defense Advanced Research Projects Agency (DARPA) in the Information Innovation Office (I2O). She provided technical and management consulting for innovative DARPA programs, which were funded at over $70 million. Her research interests lie in human-computer interaction (HCI), specifically in the design of technologies that support a range of communication and interaction needs. She is active in computer science education and STEM preparedness efforts, providing expertise for a host of funded programs funded by the National Science Foundation and the Computing Research Association, particularly those seeking to broaden participation in computer science. She also provides insight and research into the layers of intersectionality that affect Black women and girls in computer science. Jamika holds a Ph.D. in computer science, with a focus on HCI from Virginia Polytechnic Institute and State University, where she was an IBM Research Fellow. Dr. Burge is also founder and principal of Design & Technology Concepts (DTC), LLC, where she focuses on computer science design and education research. To date, DTC has consulted for Google, the National Center for Women in Information Technology (NCWIT), and the American Association of College & Universities (AAC&U), among other organizations. Her career has also included positions across academic (Spelman College and Howard University), non-profit (Smarter Balanced at UCLA), and industry (IBM Research) sectors.

Frances Colón is the former Deputy Science and Technology Adviser to the Secretary of State at the U.S. State Department. As a science diplomat in Washington, D.C., from 2012 to 2017, Colón led the re-engagement of scientific collaboration with Cuban scientists and coordinated climate change policy for the Energy and Climate Partnership of the Americas announced by President Obama. She earned her Ph.D. in neuroscience in 2004 from Brandeis University and her B.S. in biology in 1997 from the University of Puerto Rico. She currently specializes in advising local and national-level governments on science policy and evidence-based decision making. Colón is a 2018-2019 New Voices Fellow of the National Academies of Sciences, Engineering, and Medicine and a 2019 Open Society Foundations Leadership in Government Fellow. Her South Florida Climate Justice Project leverages her citizen appointment on the City of Miami Sea Level Rise Committee to create awareness and catalyze policy action that will counter the impacts of climate change and gentrification on vulnerable communities of South Florida.

Sarah EchoHawk is a citizen of the Pawnee Nation of Oklahoma, and has been working on behalf of Native people for nearly 25 years. She has been the CEO of the American Indian Science and Engineering Society (AISES) since 2013. Prior to joining in AISES, Ms. EchoHawk was the Executive Vice President at First Nations Development Institute. Previously, she also worked for the American Indian College Fund, and as an adjunct professor of Native American studies at Metro State University of Denver. Ms. EchoHawk has served on several boards and committees for multiple organizations and initiatives including the American Indian Policy Institute, Last Mile Education Fund, National Girls Collaborative, Native Americans in Philanthropy, Native Ways Federation, Red Feather Development Group, and Women of Color in Computing Research. She also serves as Principal Investigator (PI)/Co-PI on multiple National Science Foundation grant-funded projects. Ms. EchoHawk has a master of nonprofit management from Regis University and a bachelor of arts in political science and Native American studies from Metro State University of Denver.

Elena Fuentes-Afflick is Professor and Vice Chair of Pediatrics, Chief of Pediatrics at the Zuckerberg San Francisco General Hospital, and Vice Dean for Academic Affairs in the School of Medicine, University of California San Francisco. Dr. Fuentes-Afflick's research focuses on health disparities in perinatal and pediatric health outcomes. The majority of her research has focused on the surprisingly favorable perinatal outcomes among immigrant Latina women, an "epidemiologic paradox." She has also investigated the role of acculturation and immigration status on access to care and perinatal outcomes and the effect of acculturation on body mass outcomes in Latinos. Recent areas of investigation include professionalism, faculty misconduct, and academic affairs. Since 2012, Dr. Fuentes-Afflick has been responsible for overseeing all academic affairs in the School of Medicine, including the recruitment, development, and advancement of a diversified academic workforce. She is also responsible for overseeing innovative programs for faculty orientation, career development, and leadership training. Dr. Fuentes-Afflick was elected to the National Academy of Medicine in 2010, and has had extensive committee service on several National Research Council committees continuously since 2011 as well as service on the Board on Children, Youth, and Families. She received her medical degree from the University of Michigan, her master's in public health from the University of California, Berkeley, and her bachelor's degree in biomedical science from the University of Michigan.

Ann Q. Gates is Professor and Chair of the Computer Science Department at the University of Texas at El Paso. Her areas of research are in software engineering and cyberinfrastructure with an emphasis on workflows, ontologies, and formal software specification. Gates directs the National Science Foundation-funded Cyber-ShARE Center that focuses on developing and sharing resources through cyber-infrastructure to advance research and education in science. She was a founding member of the NSF Advisory Committee for Cyber-infrastructure. Gates served on the IEEE-Computer Society (IEEE-CS) Board of Governors, 2004-2009. In addition, she chairs the IEEE-CS Educational Activity Board's Committee of Diversity and External Activities and has established a model for specialized student chapters focused on leadership, entrepreneurship, and professional development. She is a member of the Computer Science Accreditation Board (2011-2013). Gates leads the Computing Alliance for Hispanic-Serving Institutions (CAHSI) and is a founding member of the National Center for Women in Information Technology (NCWIT).

Shawndra Hill is a Principal Scientist at Facebook in the Core Science Data Team in New York City and a Senior Lecturer in Marketing at Columbia University. She was previously a Senior Researcher in the Computational Social Science Group at Microsoft Research NYC. Prior to joining Microsoft, she was on the faculty of the Operations and Information Management Department at the Wharton School of the University of Pennsylvania, where she was an Annenberg Public Policy Center Distinguished Research Fellow, a Wharton Customer Analytics Initiative Senior Fellow, and a core member of the Penn Social Media and Health Innovation Lab and the Warren Center for Network and Data Sciences. Generally, she studies data mining, machine learning, and statistical relational learning and their alignment with business problems. Specifically, Hill researches the value to companies of mining data on how consumers interact with each other on online platforms—for targeted marketing, advertising, health, and fraud detection purposes. Her current research focuses on the interactions between TV content and online behaviors (www.thesocialtvlab.com). Her past and present industry partners include AT&T Labs Research, ClearForest, and Siemens Energy & Automation.

Maria (Mia) Ong is a Senior Research Scientist and Evaluator at TERC, a STEM education research organization in Cambridge, Massachusetts. She is also the Founder and Director of Project SEED (Science and Engineering Equity and Diversity), a social justice collaborative affiliated with The Civil Rights Project/ Proyecto Derechos Civiles at UCLA. For nearly 20 years, she has conducted empirical research focusing on women of color in higher education and careers in STEM and has led evaluation of several STEM diversity/inclusion programs. Ong's work has appeared in reports to U.S. Congress and to the U.S. Supreme Court and in journals such as *Social Problems* and *Harvard Educational Review*, and she was an invited speaker at the 2016 White House meeting on inclusive education in STEM. Between 1996 and 2000, she directed an undergraduate physics program for minorities and women at U.C. Berkeley; for this work, she was a co-recipient of a U.S. Presidential Award for Excellence in Science, Mathematics, and Engineering Mentoring.

Manuel A. Pérez Quiñones is Professor of Software and Information Systems at the University of North Carolina at Charlotte (UNCC). His research interests include personal information management and diversity issues in computing. He holds a D.Sc. from The George Washington University and a B.A. and M.S. from Ball State University. Before joining UNCC, he worked at Virginia Tech, University of Puerto Rico-Mayaguez, Visiting Professor at the US Naval Academy, and as a Computer Scientist at the Naval Research Lab. He is an NSF Career awardee, a Distinguished Member of the ACM, and has been recognized with the Richard A. Tapia Achievement Award for Scientific Scholarship, Civic Science and Diversifying Computing, and the CRA Nico A. Haberman award. He is originally from San Juan, Puerto Rico.

Karl W. Reid is the senior vice provost and Chief Inclusion Officer at Northeastern University and Professor of the Practice in the Graduate School of Education. Prior to joining Northeastern, he served for seven years as the executive director of the National Society of Black Engineers (NSBE), marking his return to the organization that gave him his first major leadership experience, 32 years earlier. A certified diversity professional, Dr. Reid is a leading advocate for increasing college access, opportunity, and success for low-income and minority youth. He is the author of "Working Smarter, Not Just Harder: Three Sensible Strategies for Succeeding in College...and Life." Reid came to NSBE from the United Negro College Fund (UNCF), where he oversaw new program development, research, and capacity building for the organization's 37 historically Black colleges and universities and held the title of senior vice president for research, innovation and member college engagement. Before his service at UNCF, he worked in positions of progressive responsibility to increase diversity at his alma mater, the Massachusetts Institute of Technology (MIT), which he left as associate dean of undergraduate education and Director of the Office of Minority Education. While working at MIT as Director of Engineering Outreach Programs, Reid earned his doctor of education degree at Harvard University.

Allison Scott is the Chief Executive Officer of the Kapor Foundation, an organization with a mission to increase racial diversity in tech and entrepreneurship. At the Foundation, Dr. Scott leads efforts to (a) conduct research on barriers and solutions to racial inequality in tech, (b) operate programs and invest in pathways into the tech/entrepreneurship workforce, and (c) work in partnership with stakeholders to advocate for transformational change in policies and practices to expand racial equity in technology. Dr. Scott is currently a Principal Investigator on multiple national grants to expand equity in computer science education and increase participation of women of color across the computing pipeline and is co-director for the Computer Science for California coalition, which aims to increase access and equity in K-12 CS education across the state. In her previous role as the Chief Research Officer, Dr. Scott authored foundational research on disparities in tech and entrepreneurship, inequity in K-12 CS education, and interventions to improve STEM outcomes for students of color and girls/women of color. Previous positions include Chief Research Officer at the Kapor Center; Program Leader for the National Institutes of Health's Enhancing the Diversity of the Biomedical Workforce Initiative; Director of Research and Evaluation for the Level Playing Field Institute, and Data Analyst for the Education Trust-West. Dr. Scott holds a Ph.D. in education from the University of California, Berkeley and a bachelor's degree in psychology from Hampton University.

Kimberly A. Scott is a professor of women and gender studies in the School of Social Transformation at Arizona State University (ASU) and the founding executive director of ASU's Center for Gender Equity in Science and Technology. The center is a one-of-a-kind research unit focused on exploring, identifying, and creating innovative scholarships about underrepresented women and girls in STEM. Having written and successfully raised millions in grant funding to support research about and programs for women and girls of color in STEM, Scott was named in 2014 as a White House Champion of Change for STEM Access. In 2018, Scott was invited to join the NSF STEM Education Advisory Panel created to encourage U.S. scientific and technological innovations in education in consultation with the U.S. Department of Education, NASA, and NOAA. Center projects include the NSF-funded COMPUGIRLS; U.S. Department of Education-funded COMPUGIRLS Remixed; Gates-funded project on African American Families and Technology Use; and Pivotal-funded Women of Color in Computing Research Collaborative. With four published books, the most recent is *COMPUGIRLS: How Girls of Color Find and Define Themselves in the Digital Age*, published by University of Illinois Press.

Cynthia Winston-Proctor is a narrative personality psychologist. She is also professor of psychology at Howard University, Principal of Winston Synergy, L.L.C., and Co-Principal Investigator of the NSF HU ADVANCE-Institutional

Transformation Initiative. Her academic and practice work focuses on the psychology of success of women within academic and corporate environments. Winston-Proctor also translates research to develop culturally relevant psychological science, research design and analysis, computational thinking, and behavioral cybersecurity education models for middle school, high school, and undergraduate learning environments. She is on the Editorial Board of the American Psychological Association journal, *Qualitative Psychology*, President of the Society of STEM Women of Color, and Vice Chair of the Board of Howard University Middle School of Mathematics and Science. She received the NSF CAREER Award for her early career work on the psychology of success and the meaning of race within the lives of African American scientists and engineers.